Animal Enthusiasms

Animal Enthusiasms explores how human–animal relationships are conceived, developed, and carried out in rural Pakistani Muslim society through an examination of practices such as pigeon flying, cockfighting, and dogfighting.

Based on two years of ethnographic fieldwork carried out between 2008 and 2018 in rural South Punjab, the book examines the crucial cultural concept of *shauq* (enthusiasm) and provides critical insight into the changing ways of life in contemporary Pakistan. It tracks the relationships between men mediated by non-human animals and discusses how such relationships in rural areas are coded in complex ways. The chapters draw on debates around transformations of animal activities over time, the changing forms of human–animal intimacy and their impact on familial relationships, and rural Punjabi values attached to the performance of masculine honour.

The book will be of interest to scholars of anthropology, multi-species ethnography, gender and masculinity studies, and South Asian studies.

Muhammad A. Kavesh is affiliated with the School of Culture, History, and Language at the Australian National University in Canberra, where he received his PhD. He is a recipient of the Australian Anthropological Society's Post-Doctoral Fellowship (2020).

Multispecies Anthropology: New Ethnographies
Series Editors: Rebecca Cassidy and Garry Marvin

Human–Animal Studies has come of age. Intrinsically inter- or transdisciplinary, it has encouraged scholars from diverse disciplines to use a range of approaches and resources to explore the relationships that humans have with other animals and forms of life and how these are experienced and expressed.

Animals have traditionally appeared in anthropology as aspects of cosmological systems, essential to livelihoods including hunting, fishing, pastoralism, herding, and agriculture; significant in economic systems as wealth and for exchange; celebrated in sports and other forms of entertainment and so on. More recently, animals and other forms of life have been brought to the foreground, where they are framed as important in and of themselves, rather than as reflections of other, more important relationships between people. This series is interested in this approach and the wider questions it poses about subjectivity, representation, and anthropological theory.

Uniquely, this series focuses exclusively on monographs which are based on first-hand, sustained, ethnographic fieldwork: an approach which allows for the exploration of the intricacies and immediacies of lives with other animals. It brings together detailed accounts of how humans experience, engage with, live with, other animals, but also with plants and other living matter, generated within particular social and cultural worlds as they are captured by fieldwork. This format will enable authors to pursue some of the most important questions about our lives with others – and the role that anthropology might play in our futures together.

Animal Enthusiasms
Life Beyond Cage and Leash in Rural Pakistan
Muhammad A. Kavesh

www.routledge.com/Multispecies-Anthropology/book-series/MANE

Animal Enthusiasms

Life Beyond Cage and Leash in
Rural Pakistan

Muhammad A. Kavesh

Routledge
Taylor & Francis Group

LONDON AND NEW YORK

First published 2021
by Routledge
2 Park Square, Milton Park, Abingdon, Oxon OX14 4RN

and by Routledge
52 Vanderbilt Avenue, New York, NY 10017

Routledge is an imprint of the Taylor & Francis Group, an informa business

British Library Cataloguing-in-Publication Data
A catalogue record for this book is available from the British Library

Library of Congress Cataloging-in-Publication Data
Names: Kavesh, Muhammad A., author.
Title: Animal enthusiasms : life beyond cage and leash in rural
Pakistan / Muhammad A. Kavesh.
Description: Abingdon, Oxon ; New York, NY : Routledge, 2020. |
Series: Multispecies anthropology: new ethnographies |
Includes bibliographical references and index.
Identifiers: LCCN 2020037735 (print) | LCCN 2020037736 (ebook) |
ISBN 9780367859534 (hardback) | ISBN 9781003015963 (ebook)
Subjects: LCSH: Human-animal relationships–Pakistan.
Classification: LCC QL85 .K39 2020 (print) |
LCC QL85 (ebook) | DDC 591.95491–dc23
LC record available at https://lccn.loc.gov/2020037735
LC ebook record available at https://lccn.loc.gov/2020037736

ISBN: 978-0-367-85953-4 (hbk)
ISBN: 978-1-003-01596-3 (ebk)

Typeset in Times New Roman
by Newgen Publishing UK

To my late Abu Ji
I wish I had known you

Contents

Figures

Acknowledgements

This book would not have been possible had it not been for the enormous support and kindness of all the South Punjabi animal keepers and their families. I thank them warmly for their hospitality, and for allowing me to travel with them over the course of my fieldwork while sharing ideas and opinions. For ethical, legal, and cultural reasons, I use pseudonyms for most animal keepers and try to hide the identity of their families. Still, the detailed ethnographic description I present is not entirely cryptic, and I request the reader to share in the ethical responsibility of respecting the privacy of local people. My greatest debt is to two female Research Assistants (RAs) who helped me gather the perspectives of South Punjabi women. Although I cannot name them, I want to acknowledge their determination, hard work, and superb assistance. Many people in the village helped me meet the animal keepers and accompanied me to different animal competitions. For this I thank Muhammad Ajmal Phull, Junaid Iqbal, Mian Khalid, Nazar Waqas, Agha Hussnain, Mian Ayaz, Homayun Phull, Muhammad Mahfooz, late Muhammad Masood, Nadeem Khalid, Waqas Shoaib, Maqsood Ahmad, Murad Joiya, Milad Joiya, Mahar Ghullam Abbas, Hafiz Masood, Afzal Phull, Faiz Amin, Asif Raza, Khuda Bux, and Khizar Hayat. I am also grateful to the staff of two Non-Government Organisations, police officers, lawyers, veterinarians, politicians, and others at my research site who were kind enough to offer me their time and shared their views.

At the Australian National University, my greatest debt is to my teacher and advisor Assa Doron for his kindness, critical insight, and for the generous amount of time and attention he was able to give to this project. I am grateful for his patience with my earlier drafts and for his advice on personal and professional matters throughout the journey of completing my doctoral degree. When things became difficult for me at a personal level, it was always his kind support, understanding, and encouragement that pushed me toward the completion of this book. I am indebted to Kirin Narayan, who always had a gracious smile for me and encouraged me to add beauty to my writing. While attending her course on "Ethnographic Writing" and co-authoring an article on *shauk*, I experienced the possibility of feeling alive in writing. It was in her house, in the company of her two beautiful cats, Chunni and Munni, that I started reworking my PhD dissertation into a book manuscript. I am also greatly indebted to Natasha Fijn who, as a friend and an advisor, opened the door for me to understand the more-than-human world. Her insightful comments, excellent guidance, and suggestions have been greatly valuable to this project.

I would like to show my gratitude to other people who provided their useful comments on earlier drafts of this project. I am particularly thankful to Matt Tomlinson, Philip Taylor, Kuntala Lahir-Dutt, Andrew McWilliam, Andrew Kipnis, Alan Rumsey, Lacy Pejcinovic, Thuy Do, Maxine McArthur, and Owen Bullock for their insights into earlier versions. I am grateful to Kenneth George, Dipesh Chakrabarty, Robin Jeffrey, and Jürgen Wasim Frembgen for suggesting useful literature for this project, and to Mark Mosko, Kathy Robinson, Shameem Black, Meera Ashar, Margaret Jolly, Trang Ta, and many friends and colleagues for offering valuable advice during seminars. I am indebted to Naisargi Dave, Tanya Jakimow, Eva Nisa, Joseph Alter, David Howes, Laurian Bowles, Beth Uzwiak, Fouzieyha Towghi, Keith Barney, Luca Tacconi, Patrick Guinness, Kama Maclean, Tom Cliff, Caroline Schuster, Francesca Merlan, and David Macdougall for their enormous kindness. I also thank Ursula Sims-Williams from the British Library, and the staff of the SOAS library for their assistance in scanning and sending rare Indian manuscripts on pigeons. I am obliged to Ms Karina Pelling and her CartoGIS team at the Australian National University for producing maps for this project. I am highly thankful to Samina Choonara who, as an excellent scholar of Pakistan's history and culture, provided her highly critical comments and important edits.

I would like to express my deep gratitude to the Australian Government's Endeavour Award for its generous financial support for my PhD project. The School of Culture, History, and Language at the Australian National University helped me in various ways to meet my study expenses, sometimes offering a fee waiver and fieldwork support grant, and at other times providing assistance in obtaining rare books from international libraries. I am highly indebted to the School Director, Simon Haberle, the School HDR administrators, Jo Bushby and Etsuko Mason, and other staff of the School, including Kirsten Farrell, Gouri Banerji, Joshua Burgess for always providing timely support. I would also like to acknowledge other grants that helped me at different stages of this project, these include: Australian Anthropological Society Postdoctoral Fellowship, Raymond Firth Award by the Australian National University, South Asian Research Institute's (SARI) Student Support, Robin Wood Award by the Australian Anthropological Society, ANU Student Leader Scholarship, ANU Postgraduate and Research Students' Associations' Award, Quaid-i-Azam University Student Aid, and the Australian Network of Student Anthropologist's Conference Support Grant.

I have presented parts of the book at many conferences, seminars, and workshops over the last few years and have received helpful criticism and comments from a wide variety of scholars at Wollongong University,

University of Sydney, University of Adelaide, Fatima Jinnah University Rawalpindi, and at various colleges in the Australian National University. Parts of the book have been published; an earlier draft of Chapter 1 appeared as "From the Passions of Kings to the Pastimes of the People: Pigeon Flying, Cockfighting, and Dogfighting in South Asia" in *Pakistan Journal of Historical Studies*, and portions of Chapter 4 were published as "Dog Fighting: Performing Masculinity in Rural South Punjab, Pakistan" in *Society & Animals*, and as "From Colony to Post-Colony: Animal Baiting and Religious Festivals in South Punjab, Pakistan" in David W. Kim's edited book *Colonial Transformation and Asian Religions in Modern History*, CSP Newcastle.

I wish to thank Rebecca Cassidy, Garry Marvin, and Radhika Govindrajan, the editors of the book series *Multispecies Anthropology: New Ethnographies*, for their close reading, comments, and support. This book has also benefitted immensely from the extremely valuable and insightful comments from reviewers. I am also grateful to my editor, Katherine Ong at Routledge for her expert advice, and to Kangan Gupta, for her tremendous support.

My friends in Australia have helped me with their encouragement, ideas, and support on many occasions. I am grateful to Syed Shah Faisal and Daisy Bhabi, Abdul Razaque Channa, Muhammad Adeel, Tayyaba Malik, Zaime Bujor, Mukesh Tiwari, Milad Ketab, Omar Pidani, Fitrilailah Mokui, Neal Chen, Sarita Dhounchak, Colum Graham, Helen Abbott, Joyce Das, Umar Assegaf, Myra Mentari, Erlin Erlina, Sacha Cody, Fina Itriyati, Nita Sebastian, Ruhan Shahrir, Mayank Khanna, Sherman Tan, Isabela Burgher, Sana Ashraf, Eleonora Quijada, Shamim Homayun, Mariko Yoshida, Benny Tong, Khurram Aftab Kiani, Zoe Hatten, Valerie Mashman, Benny Tong, Gita Nasution, Xeem Noor, Karen Hansen, Maria Ortega, Kimlong Chheng, Elvin Xing, Mizan Rahman, Tony Jefferies, Charlotte van Tongeren, Lina Koleilat, Justin Lau, Annie McCarthy, Bethune Carmichael, and Mark Jones for helping me in countless ways. Some friends read parts of this book and provided invaluable comments. I would like to offer my gratitude to Alexandra McEwan, Kathleen Varvaro, Simon Theobald, Julia Brown, Nonie Tuxen, Jo Thurman, Ben Hegarty, Justine Chambers, Poonnatree Jiaviriyaboonya, Said Alvi, and Fay Styman for thoughtful and constructive feedback on my writing.

I am extremely grateful to people in Pakistan for their generosity and support. My friends Ishaq Isra, Salman Ijaz, Muhammad Sadiq, Aurangzaib Buzdar, Safdar Klasra, and my college lecturer Mukhtyar Shah inspired the interest in studying Saraiki culture. Sayed Tahir Parveez, Hafeez-ur-Rehman, Waheed Iqbal Chowdhry, Aneela Sultana, Meena Zulfiqar, Sufyan Abid, and especially my graduate supervisor, Tariq Mehmood, encouraged me to explore anthropology further. I would also like to thank other people at the Department of Anthropology in the Quaid-i-Azam University, including Anwar Shaheen, Sadia Abid, Rao Nadeem, Inaam Laghari, Anwaar Mohuddin, Sajjad Haidar, Ilyas Bhatti, Waqas Saleem, Ikram Badshah, Muhammad Sajjad librarian, Rafique Sahib, and Tabbasum Rehmani for their continuous kindness and support. I cannot thank my friend and mentor Muhammad Bilal enough for encouraging and supporting me in so many ways. His advice and guidance has always been a source of intellectual and personal nourishment. Other friends in Islamabad, Imad Mastan Khan, Ajmal Khan Kakar, Salman Shah, Zubair Ahmad, Tariq Rahim Shah, Aisha Farooq, Muhammad Ali Durrani, Ambrin Sharif, Habbat Kibzai, Rashid Kazmi, Mehr un-Nisa, Adnan Jameel, Hassan Mahar, Ghazanfar Ilyas, Muhammad Majid, Muzammal Bilal, Nizam ud-Din, Maria Ali, Waqas Badshah, Nadia Khaliq, Asim Hussain, Javeed Bhutto, Hamasa Tariq, Akram Ali, KD Baloch, Raza Hayder, Mahar Akmal, Rabia Naseer, Mumreez Khan, Mehwish Rasheed, Abida Zafar, Awon Muhammad, Afreen Athar, Sherbaz Khan Baloch, Ahmad Raza, Sara Yousaf, Qamar Zaman Bodla, Saiqa Ashraf, Umar Draz Virak, Adnan Mughal, and Rehman Malik have been a source of great ideas and inspiration.

My family has been a source of endless love, support, and encouragement, particularly my parents-in-law, Shehzad Akhtar Raja and Nazneen Shehzad. My mother, as a single parent, has always provided her care, love, and affection, and her presence in my life is a blessing and a source of motivation. I owe a great deal to my brother, Arshad, who raised me like a father since I was two years old, and ensured that I receive the best education possible despite financial troubles and mounting debt after the passing away of our father. I would also like to thank my other siblings for their love and sustained support.

My lovely wife, Zara, has supported me in so many ways. She patiently read many drafts of all the chapters, gave me constant feedback, and always pushed me to explore my limits. As a friend, she has provided companionship and advice on a daily basis and helped me overcome numerous difficulties. My daughters Serosh Zaynab and Hoorain Aisha are a blessing for me and keep me going. Their every breath is a source of comfort to me and their every act brings great joy. I want to live and die watching you healthy, happy, and shining.

Notes on translation and transliteration

Seraiki is the most commonly spoken language in South Punjab, while Mewati is the second largest spoken language of the area, a dialect of Urdu spoken by those who migrated from India at the time of the Partition in 1947. In my transliteration, I have tried to write the words as I heard them, sometimes in Seraiki and sometimes in Mewati. All local words are italicised, whereas the diacritical marking follows standard conversions of The Library of Congress's Urdu Romanisation method (2013). I have decided against the use of diacritics for people's names. Also, for the ease of English speaking readers, I have used the plural form of local terms by adding an "*s*" at the end (such as *ustāds, melas, shareeks*). Other than some major consonants, for instance *c* as "ch" in chair, the scheme of transliteration is as follows.

a as in shut   *ā* as in hard
e as in met   *y* as in say
i as in sit   *ī* as in machine
n as in nose   *ṇ* as in tongue
u as in pull   *ū* as oo in food

For other consonants, the main differences are:

d is softer as *th* in gather
ḍ is harder as in diary
t is softer as in teeth
ṭ is harder as in true

Similarly, *ṇ* and *ṛ* have a retroflex flap.

Introduction

"I've heard that white people have no *izzat* (honour). Is that true?" Raza asked me one fall morning in 2008. I had just begun a six-month-long ethnographic fieldwork as part of the requirement for my MA thesis in a remote South Punjabi village, studying the passion of pigeon flying. At the time, I found it disconcerting how many pigeon flyers seemed reluctant to share the details of their work with me, and mostly spoke to each other in coded language, strategically deployed to exclude me from their conversations. And yet, Raza was kind enough to introduce me to the complex art of pigeon flying, explaining the practice and its associated vocabulary. The reason for my exclusion from other flyers of the area was simple: I was different in almost every way. I was unmarried and in my 20s, whereas most pigeon flyers were married men with greying hair. I had come from Islamabad, the federal capital, which was a place foreign to most villagers. Above all, I was there to "study" their passion of pigeon flying, an activity that carried a strong social stigma and many flyers felt hesitant to talk about it. Even when granted a visit to a rooftop with a pigeon flyer, the sense I got was that the responses to my questions were typically short and snappy, again in what was an unintelligible pigeon-flying vocabulary. This ongoing rebuke continued for several weeks until I met Raza, an unmarried man of my age. Sensing my curiosity and persistent interest in pigeon flying, Raza decided to take me up to his rooftop where he kept about 100 pigeons of all colours. There, he patiently answered many of my queries. When I was done, Raza moved on to offer a few observations of his own about the nature of modern life. "Look, they (white people) live together without marriage, lie on the beaches with hardly any clothes on, and even drink alcohol with women. How can a man drink with a woman? They have no *izzat* at all." Opening the doors of his pigeon coops, he continued, "They don't even have a word for *izzat* in English … the concept does not exist for them." Considering my knowledge of English, tertiary education, and urban background, he looked at me for answers.

In search of a more authoritative response, I suggested we check the Urdu to English dictionary on Raza's new Nokia phone. There we found that the English translation of *izzat* covered several words: "honour, prestige, pride, reputation," yet Raza remained adamant that these words were an approximation rather than an accurate translation. "They never think about *izzat* like we do here in Pakistan," he suggested, while looking at his cherished pigeons who fluttered out of the coop, cooing with pleasure and perching on Raza's arms and shoulders. By now, the pigeons appeared ready for a short flight before their first feed. "These pigeons are my *izzat*," declared Raza, "when they fly together in a cluster, other flyers of the village praise me and appreciate my mastery." Waving the *dogaza*, the vertical net made with two mulberry sticks, Raza commanded his pigeons to fly and as the birds ascended in the air, Raza seemed to have disconnected from his surroundings, freed of his earthly body, and oblivious to the world and its gravitational concerns. This was especially remarkable to see, because Raza, like all other pigeon flyers of the area, was constantly castigated for his absorption with pigeons. This embodied passion that fuelled his enthusiasm and endowed him with *izzat* among other flyers, was also the source of his social disgrace (*beztī*). His relatives, neighbours, and even his close family members continuously denounced him for flying pigeons: a practice they considered both immoral and self-indulgent. While I lived with him and later with other animal keepers of the village, joined in their daily activities, and often had long discussions around various issues, the term *izzat* kept coming up quite often in our conversations. It seemed to be the central idea that shaped men's social lives and their everyday relationships with each other. It was also what influenced their preferences, attitudes, and knowledge of self and non-human others. How can we conceptualise the idea of *izzat*? Why and how, as Raza suggests, is it different from the English word "honour"? And how does it coincide with the idea of disgrace to shape everyday life choices of people living in rural South Punjab?

Izzat derives from the Arabic root word "*izza*," meaning "might, power, strength; honour, glory; high rank, fame, renown; self-esteem; pride" (Wehr 1979, 713). In its different forms and pronunciations, it is widely used in Persian, Urdu, Nepali, and Hindi. More specifically, in the context of Pakistani Punjab, anthropologists have translated *izzat* as "honour" (Lyon and Fischer 2006, 48); and considered it of the utmost importance in the indigenous value system (Frembgen 2006, 248) that influences an individual's identity and status (Lefebvre 1999, 264). Many researchers have found the concept of *izzat* entangled with ideas of control. To them, a person who controls herself/himself (behaviour, emotions, feelings, bodily practices) is viewed as possessing *izzat* (Lyon 2004, 14–21; Fischer 2006, 331). In addition to self-control, according to Fischer (2006, 331; 1991, 108), the *izzat-dar* (honourable) person should exert control and influence over other members of the community through, for instance, practising family and community values such as keeping women in *purdah*, having obedient sons, resolving village conflicts, joining committees, and heading political groups. In my experience of working in rural South Punjab over a decade, I found that the term *izzat* covers a range of ideas, from the control of financial resources to correct social conduct and bodily comportment.

Izzat, as Mark Liechty (2003, 83–85) suggests in his writing about middle-class consumerism in the context of Nepal, is an "unbelievably powerful force" that many people try to command by maintaining a privileged *moral* and *material* high ground in society. It is a part of a person's habitus, something that Pierre Bourdieu (1977, 81) regards as embodied competence, expressed in a series of internalised modes of behaviour, tastes, and appearance that render social interaction possible. For instance, in rural Pakistan *izzat* may influence career aspirations, marriage choice, voting practice, and even religious preference of selecting one mosque over another for daily worship. It inspires some people to host guests, engage in charity, or join a pilgrimage to Mecca. It shapes the practice of agnatic rivalry, endogamy, exchange marriage, and even "honour killing." Most people are expected to embody this ideology and demonstrate it regularly through their actions, choices, and practices. It therefore seemed reasonable for Raza to argue that people in the West, about whom he learned from Hindi-dubbed Hollywood movies, have "no *izzat* at all." The idea of honour, as conceived and experienced in his life, to him seemed absent from western societies. *Izzat* was, therefore, a critical concept that encapsulated his present outlook and was an embodiment of his personal history, fear, and ambitions, mediated through a range of institutions and social relations.

Raza was born into a privileged, landowning family. His grandfather owned 500 acres of valuable farmland that he shared among his four sons. In his desire for a male offspring, Raza's father, third eldest among the siblings, ended up having six daughters in a row, spaced barely a year apart. Desperate for a son, he married a second time and Raza was born to him. However, after the death of his grandfather, Raza's father started selling the inherited farmland and investing in maintaining a lavish lifestyle. In less than fifteen years, all four brothers had sold off their inherited land to purchase new cars and employ chauffeurs, build new houses, prepare rich dowries for their daughters, and spend extravagantly on celebrating the annual death anniversary of their father. Such practices raised their social status, however, in rural South Punjab where agricultural land is the primary source of economic standing, this lifestyle could not be sustained for long. Over the years, as the farmland shrank, so did their *izzat* in the village.

In the early 2000s, young Raza witnessed these transformations. Although people did not dare say anything to his face, he noticed many of them snickering behind his back. In his village, family honour was tied to the possession of and control over agricultural land and resources. Selling ancestral farmland was like severing one's roots. In 2004 when Raza's father was diagnosed with bone cancer, the remaining farmland was sold for his treatment in Lahore. Raza went to school but he had to leave it to care for his ailing father and attend to the daily needs of his two mothers and unmarried sister. It was in this gloomy atmosphere that Raza found a passion for pigeons. Initially, he brought two pigeons home from a friend, and then the number of birds grew at great pace. He told me it was the company of pigeons that gave him respite from his troubles. Seeing his pigeons flying freely, innocent of all social worries, helped him develop a renewed sense of optimism. And yet, no one in his family or neighbourhood understood his enthusiasm; instead they viewed it as a frivolous distraction. People expected him to work hard and earn money for the family, set up a small shop in a nearby town, or migrate to Dubai as a labourer to remit money back home. Nobody expected him to spend his days and nights in breeding, feeding, and flying pigeons.

Raza's life trajectory was shaped by the expectations and responsibilities associated with maintaining *izzat* in the demanding rural setting of Punjab. This cultural concept implicitly and explicitly influences people's attitudes, choices, and decisions and drives their motivations and actions. And yet, of all places, Raza found *izzat* in the company of pigeons. According to him, pigeons provided him personal delight and elevated his position among experienced flyers of the village. For him, and for many other animal keepers, *izzat* was not garnered by making astute economic choices through marriage practices, or political decisions, rather, it was achieved by developing an intimate and lasting relationship with cherished animals, showcasing one's mastery in raising them, and by winning animal competitions. Their conception of *izzat* is then something similar to what Bourdieu calls "symbolic capital," which when recognised alongside other forms of capital (economic, cultural, social), enables people to elevate their social position by distinguishing from other members of the same class (Bourdieu 1977, 180–181; 1990, 120–121). Bourdieu's original conceptualisation of symbolic capital was based on his fieldwork in Algeria and his analysis of the education system in post-WWII French society. The context and scale are apparently quite different from my study of rural South Punjabi society. However, as over years Bourdieu develops this concept as part of a wider theoretical framework around theory of practice, I find his notions useful in understanding animal keepers' interpretation of *izzat* which affects their social standing and influences their role and behaviour in the *field* of keeping and competing animals.

In rural Pakistan, men's honour (*izzat*) is interwoven with their performance of masculinity (*mardāneat*). In most cases, only men are supposed to possess *izzat* which is closely tied to strict control over the bodily actions and behaviour of the women of their family. Any "dishonourable" and "shameful" act by a woman, such as not wearing an appropriate form of *purdah*, dancing, falling in love, marriage of choice, elopement, or even talking to strange men on the phone can result in the loss of *izzat* of her husband, father, brother, and sometimes, the male members of the extended family. Thus, *izzat* in rural Punjab is not only about being a good man but also, as Michael Herzfeld (1985) observes in the context of Greek, being good at being a man. Herzfeld (1985, 16) considers masculinity as "*performative excellence*, the ability to foreground manhood by means of deeds that strikingly 'speak for themselves'" (original emphasis). Such explanations resonate with my own observations in the field where masculinity was an inherently performative act. For rural Pakistani animal keepers, as I show throughout this book, *manliness* brings *honour* through actions. It is not simply determined by the act itself (e.g., flying pigeons), but crucially in the way the act was performed: the mastery attached to it, the skill of the man and the animal during the competition, and how distinct it was from other related actions.

By focusing on more-than-human sociality, I explore the knotted relationship between honour and masculinity in rural Pakistan. Broadly, I conceptualise masculinity as it is defined by Raewyn Connell, as "a place in gender relations, the practices through which men and women engage that place in gender, and the effects of these practices in bodily experiences, personality and culture" (1995, 71). Yet, Connell is equally sensitive to the more nuanced aspects of masculinity, arguing for the need to recognise the plurality and multiplicity of masculinities. Taking a lead from Connell's idea of *multiple masculinities*, I propose that masculinity in rural Pakistan is subtle and complicated, rather than one-dimensional "hyper-masculinity," and a close understanding of this multiplicity lead us to concretely conceptualise the category of *izzat*. As I try to untangle the relationship between men and non-human animals through various phases of breeding, feeding, decoration, training, and competing, I ask what aspects of masculinity are most prominent when the men showcase behavioural, emotional, and bodily control. Why do the men cling to some dominant masculine qualities, like courage (*jur'at*), bravery (*dilerī*), and strength (*zor*), and how are such attributes infused in the seemingly non-masculine activity of pigeon flying? By exploring the cultural debates over honour and masculinity, the book examines such questions and shows how a "true man" is expected to perform and bring meaning to his social life which, in most cases, is inclusive of indifference and involves the exploitation of personal emotions.

The enthusiasm for animals

To a great extent, this book is inspired by my own enthusiasm, an affection I share with Nikola Tesla, if one is to imagine a completely different time and place. In room 3327 of Hotel New Yorker, the great inventor Tesla is said to have spent much of the time during the last decade of his life feeding pigeons. When *The New York Times* reporters asked Tesla about his thoughts on marriage, he replied with a story about a white pigeon with much affection. "She understood me and I understood her ... Yes, I loved her as a man loves a woman, and she loved me." Tesla stated, "That pigeon was the joy of my life. If she needed me, nothing else mattered. As long as I had her, there was a purpose in my life" (Cheney 1981, 228). Unlike Tesla, I never had a chance to keep pigeons in my house in a small South Punjabi village. Yet, I clearly remember how this dervish-like bird first gripped my attention in my childhood years.

At the age of nine years, one hot summer afternoon as I cycled home from the village school, a vigorous stroke on the pedal snapped the fragile chain and I was forced to continue on foot. After dragging the bike and the heavy bag full of thick books for some time, I managed to reach a roadside cycle mechanic, Akmal, who was on the afternoon prayer break. In the tremendous heat, the only respite was the village tube well, fiercely pumping out water to quench the thirst of the vast cotton field. As I sat there under the banyan tree near that tube well waiting for Akmal to reopen his shop, a beautiful red pigeon alighted from seemingly nowhere.

Initially, I stared at her and she stared back. Slowly, I approached her, and then (as we humans do) tried to catch her. Each time I tried, she flew a couple of feet away from me cooing softly and playfully, as though coaxing me to try again. Maybe she had come looking for grain or possibly was just curious at the sight of a stranger near the well, or maybe it was her way of initiating friendship across the boundary of species. After many attempts, the pigeon remained elusive but the memory stayed with me. From that time, I began noticing the wondrous world of pigeons and pigeon flyers, including the strange whistling that accompanied the sound of fluttering wings. This multi-pitched whistling, I soon realised, was the specialised way of some skilful men to direct the flight of their trained pigeons. However, because of the local stigma attached to pigeon flyers in the area, my family refused to let me keep pigeons or to ever meet a flyer; instead, I was sent 300 kilometres away to a boarding school. Yet, pigeons always occupied my imagination, perhaps because they represented the pastoral setting of my early years. Many years later, I found myself making my way back to these mystic birds on a six-month long ethnographic study of pigeon flying as part of my MA (2009) thesis.

This book is also about the world of fiercer human–animal relations; one where fighting and bloodshed dominate. In October 2014, when I left Australia for Pakistan for my PhD fieldwork in rural South Punjab, I chose cockfighting and dogfighting, along with pigeon flying, as topics of my research with the motivation to recommend strategies to policymakers and practitioners to stop these "cruel" and "immoral practices." I planned to highlight the attitude of state and local authorities to "animal sports," and joined the company of animal rights activists in Islamabad for a week. With these urban, educated, upper-middle-class activists who lived 500 kilometres away from rural South Punjab, I discussed the ethical and moral nature of these activities where animals were deployed to amuse their caretakers, and sometimes incurred severe or even fatal injuries. Many activists were convinced that it was the rural South Punjabi men's illiteracy, poverty, and backwardness that attracted them to "blood sports."

Yet, once I arrived in rural South Punjab and met with cockfighters and dogfighters and learned about their lives, a very different worldview was revealed to me. These animal keepers never referred to their chosen pursuit as "blood sport" or even "sport"; rather, to them it was part of their personal enthusiasm, affection, and passion. Men who fought roosters and dogs considered these animals interactive subjects, communicating companions, and intimate partners with whom they developed mutual affection and deep sociality. They did not see their animals as fighting objects but regarded them as beings with distinctive personalities, elegant fighting skills, and courageous attitudes. To them, each animal was different and possessed unique characteristics, such as the will to fight and the ability to outmanoeuvre the opponent, either with an aggressive stance or with clever tactics. As I engaged in "participant observation" and learned different ways of whistling to fly pigeons, prepared the feed for gamecocks, and assisted dog keepers in training their dogs, I gained an understanding of what was largely an embodied, unspoken attachment to animals, inspired by the men's passion to achieve *izzat* through the performance of their *mardāneat*.

How can the practice of fighting animals be considered a passion for animals? How can we understand more-than-human sociality as a relationship of mutual affection? Scholars studying human and animal relationships emphasise the importance of animal subjectivity. Some gauge this subjectivity by realising the animals' suffering and their production as a commodity for human use (e.g., Francione 2010; Francione and Garner 2010; Bulliet 2005; Sunstein 2004; Wise 2002; Best 2009; Fraser 2009; Webster 2005), while others look at the animals' integration into human cultures, the symbolic meanings animals carry, or how they are invested with different attributes and impact on their human caretakers (e.g., Haraway 2003, 2008; Govindrajan 2018; Jerolmack 2009; Cassidy 2002; Herzog 2010; Marvin 1988; Fijn 2011; Kohn 2013; Hurn 2012; Ingold 2013). It is the second group of scholars, mostly dominated by anthropologists, whose interpretation of animals' socioculturally specific role inspired my own analysis of the patterns of more-than-human sociality in rural Pakistan.

Emerging literature in the field of multi-species anthropology, particularly the argument of "anthropology beyond human(ity)" (Kohn 2013; Ingold 2013), has helped us challenge the place of humans as the *only* subject of focus in the ethnographic description. For instance, Eduardo Kohn's (2007) "anthropology of life," suggests we study the human world "within a larger series of processes and relationships that exceed the humans" (2007, 6). Through careful analysis of the Amazonian interpretations of dog dreams, Kohn proposes cutting across the established nature/culture distinction and envisions an ethnography that is not confined to humans but is concerned with human interactions with other life forms. Similarly, Anna Tsing (2013, 27) in her discussion of "more-than-human sociality" argues that living beings other than humans should be viewed as fully social, with or without humans. Both "anthropology of life" and "more-than-human sociality" guide us to consider how anthropology as a discipline can benefit by engaging with what Eben Kirksey and Stefan Helmreich refer to as "multi-species ethnography," which explores "how a multitude of organisms' livelihoods shape and are shaped by political, economic, and cultural forces" (2010, 545).

As a work of multi-species anthropology, the book focuses on the entanglement of the human and animal lifeworlds in rural Pakistan and argues that to explore such entanglement, we must pay attention to how human and animal lives unfold through a complex relationship of care and violence.[1] Pigeon flyers, cockfighters, or dogfighters *care* for their animals by decorating their bodies and providing them protection and a good diet, and yet, since they engage them in competitive activities that may result in exhaustion, injury, or death, this appears as *violence* to urban animal rights activists. Such a knotted relatedness between animal care and violence, as Radhika Govindrajan (2018, 36–37) has argued while describing the deep emotional connection between the person performing the sacrifice and the animal being sacrificed in India, sometimes involves a history of affective labour and embodied attachment. In rural Pakistan, pigeons, roosters, and dogs also have histories of "becoming-with" humans, cultivated over years of intimacy and sociality (Haraway 2016, 15; also see Haraway 2008). Despite poverty, most of the men invest time and money on their cherished animals and provide them utmost protection from harsh weather. Most of them do not conceive of their animals as passive objects of entertainment but take them as active subjects who indulge in mutual interaction with their human keepers, enliven their lives, and communicate various meanings. However, at the time of competition when a man's masculine *izzat* is at stake before his peers, the animals are expected to perform. There in the celestial and terrestrial arenas, the skilful display of pigeons, roosters, and canines not only becomes a proof of human–animal intimacy to the audience but also to the animal keeper. This more-than-human sociality that binds human and animal lives through care and neglect, love and violence, and affection and indifference is usually encapsulated in a single, culturally meaningful word—*shauq*.

Shauq of keeping and competing animals

The word *shauq* drives from the Arabic root word *šāqa*, and is translated in English as delight and joy; to arouse longing and craving; to awaken a desire; and to long, yearn, crave, and covet ardently (Wehr 1979, 577). In South Asia, it is usually differentiated from the English word "hobby." Mark Liechty emphasises this distinction and argues that *shauq* is an individual's "innate inclination … something of more significance" than the English word hobby (2003, 81–82; see also Liechty 2010, 40). Ethnomusicologist Mark Slobin (1976, 24) also sees *shauq* "somewhere in the zone between a hobby and an obsession" and describes it as "a preoccupation with certain objects or activities, but one that does not constitute the individual's principal job or determine his station in life." Nita Kumar, in her seminal study of the artisans of Banaras in India, describes *shauq* as an "intense love for an activity for its own sake" (1989, 163).[2] In our description of *shauq*, Narayan and I (2019, 725) explain it as something invaluable that allows people to acquire specialised knowledge and achieve emotional well-being. *Shauq*-driven activities, we suggest, "can help make life worth living amid often difficult and rapidly transforming circumstances" (2019). In this book, however, I take the lead from a work of the twelfth-century Persian Sufi poet and philosopher, Muhammad Abu' Hamid Farid ud-Din Attar, to emphasise the distinct ability of *shauq* to generate more-than-human sociality, influencing social relationships, exposing fears, and guiding a person through various stages of life.

In *The Conference of the Birds* (*Manteq at-Tair* c.1177), Attar narrates the fable of the different species of birds who gather together and ponder the question of their king. They are informed by the hoopoe that they indeed have a king, the immortal Simorgh (a mythical bird from pre-Islamic Iran). The hoopoe tells them that the elusive Simorgh lives on the mysterious Mount Qaf, and the journey to him is fraught with danger. The enthusiastic birds choose the hoopoe as their leader and set out in their quest to seek Simorgh. Along the way, the birds ask the hoopoe about the perils of the journey, to which he responds with illustrative anecdotes. Sensing fears, showing inadequacies, or being attracted by worldly desires, some birds begin to leave, with only thirty of them agreeing to carry on. These thirty exhausted birds inquire about the length of the journey, and the hoopoe describes the seven valleys (or stages) they must cross to reach the Simorgh—seeking (*talab*), love (*ishq*), knowledge (*ma'rifat*), detachment (*istighna*), unity (*tawhid*), bewilderment (*hayrat*), and poverty and annihilation (*faqr o fana*). At the end of their journey, the wearied birds finally reach the court of Simorgh. There they see a mirror and their own reflection in it: these thirty birds are the Simorgh (Persian homonym for thirty [*si*] birds [*morgh*]), and the arduous journey is revealed as being an act of self-discovery.

Attar's allegory uses the birds' journey to demonstrate wayfarers' quest to find a guide (the hoopoe, in this case) who can lead them through the mystical path (or seven valleys) to reach God (or the Simorgh). Yet, I also see the journey of thirty birds as inspired by their dedicated *shauq* and a process of self-discovery. As the birds' *shauq* leads them across the seven valleys, they discover the truth about life (see also Varzi 2006, 216–217). With each valley, they discover inner secrets and reach closer to achieving a state of emotional elevation. The discussion of the seven valleys is important for understanding the notion of *shauq* and I will come to these . later in Chapter 6, however, here it is important to note that Attar's protagonist is an individual, and his focus is on *shauq* that guides the individual's path in a journey of self-exploration.

During fieldwork, whenever I asked pigeon flyers, cockfighters, and dogfighters about their passionate interest to keep, fly, and fight animals, they had a standard reply, "*ae sab shauq dī gāl he,*" or it's all about *shauq*. Shauq, they said, shapes their choices, values, and aspirations. They described it as the process of discovering their inner self in a world of social turmoil, domestic expectations, and economic responsibilities. To most of them, participation in their chosen pursuits is not a side hobby but complete dedication to experience life at its fullest. To a great deal, the rural South Punjabi animal keepers' *shauq*, borrowing an expression from Loïc Wacquant (1995, 507) from his study of Chicago boxers, makes them "what they are: it defines at once their innermost identity, their practical attachments, and everyday doings."

Through their *shauq*, many rural Pakistani animal keepers say they enter into a meditative state where they forget themselves, lose track of time and mundane commitments, and achieve an experience similar to what psychologist Mihaly Csikszentmihalyi (2014, 215–216) calls *flow*. Csikszentmihalyi describes flow as an optimal and intrinsically rewarding experience generated through deep concentration, altering the subjective experience of time and space and allowing a person to view and achieve a sense of control over life goals. He develops this notion of flow by interviewing inventors, artists, and athletes (among others) and argues that flow in any field entails a deep focus— a meditation-like state—that enables people to experience their chosen pursuit in an enormously enjoyable and enriching manner. The *shauq* of South Punjabi animal keepers resembles a flow-like experience as it harbours the capacity to experience deep joy. However, as I show in the book, the concept of flow does not sufficiently capture the societal entanglement of animal keepers with their birds and canines. Flow guides us in understanding the intensity of experience and depth of emotions, and the way these emotions generate self-interest and turn an activity into an enjoyable practice. It does not, however, examine why an enthusiast would engage with his practice beyond the objective of achieving a fulfilling experience, what societal reactions an experience of flow engenders, or in what ways flow affects a person's familial and community relations in a particular cultural context. Following Attar's discussion, I analyse these and other meanings associated with *shauq* and explore how it intertwines the quest of self-discovery with the attainment of symbolic rewards by keeping and competing animals.

In the remainder of this book, the reader will notice that *shauq* appears as a personal journey filled with enthusiastic devotion and zest to explore one's inner joy. I take it as a crucial concept that leads people towards a higher mode of being where they learn, improvise, and innovate new techniques through their daily interaction with animals. I also analyse it as a cultural category that allows people to define and imagine their personality within a wider cultural context. Many animal keepers believe that through their *shauq* they can achieve respect among peers, develop friendships, and gain "something" intangible that would otherwise be inaccessible to them. That "something" is usually embodied and experienced rather than expressed in words. *Shauq*, in this sense, becomes an intricate part of their habitus which generates and organises their practices with their animals and underlies the structuring of all major experiences of their social lives (Bourdieu 1977, 78; 1990, 53). However, *shauq* is not only implicitly adopted and explicitly inculcated; it also articulates a person's freedom, a space in which to exercise agency, and to care for the self—something that approximates what Foucault (1988, 18) termed "technologies of the self."

This book describes my engagement with the *shauq* of pigeon flying, cockfighting, and dogfighting to explore how human–animal relations are conceived, developed, and carried out in rural Pakistan. It shows how *shauq* animates the animal keepers' everyday lives, and influences their lived experiences by (re)defining their social relationships, (re)shaping their symbolic practices, and (re)ordering their ideological orientation. Animals, as I stated earlier, are not seen by most men as objects of human entertainment, instead, they are a vehicle to cultivate a person's *shauq*. As this book probes the lives of rural Pakistani animal keepers, it examines how the *shauq* of keeping, flying, and fighting animals refashions their social lives, leads them to discover their self, and enables them to win or lose masculine *izzat* within and beyond the world of competing animals.

Living with the *shauqeen*

Let me take you back to my own story. After the strange encounter with the red pigeon, I planned to keep the bird at home. However, before I could actualise my ambition, I was sent off to a famous boarding school in Muzaffargarh, a pre-cadet school that was a feeder to major military colleges of the country. After finishing eighth grade, I took entry tests to six military colleges, failing in all attempts. My desperate family now turned their attention to making me a software engineer, so I found myself at Sadiq-Egerton College, which the British had founded in the city of Bahawalpur in 1886. Unsuccessful, I moved to Multan for an undergraduate degree in sociology and then proceeded to graduate in anthropology at the Quaid-i-Azam University in Islamabad. Over these years, I identified myself as a South Punjabi even though I had spent more time in boarding schools, college hostels, and university accommodations than in my village. To the local people, I was the son of a small mango farmer, but an educated person; born in the village, but lived in the cities; attended cultural events like rural festivals, but never regularly; could converse in the local language, but had never formed associations of friendship. It was only during my MA and PhD fieldwork that I returned to the area for a substantial period. During this time, I started looking at and appreciating "local ways," with a special interest in people and their animals. This unique insider/outsider experience was especially productive in developing an anthropological imagination and in generating insights into life in rural South Punjab.

I mostly carried out fieldwork for this book in Lodhran district, and in some parts of Bahawalpur district.[3] Pakistan has four provinces, and Punjab is the largest in population, agricultural growth, and industrial productivity. Although most agricultural products (cotton, sugar cane, and mangoes) come from South Punjab, mainstream development schemes often neglect this area in comparison to central Punjab (Jamal et al. 2003, 97). The land in rural areas is mostly owned by landlords who contract their farms to low-class farmers in return for half of the annual production. Most of the animal keepers I met during my fieldwork were working as contractors for these landlords, hardly making $40 a month.

The poverty level in rural parts of Lodhran and Bahawalpur is very high. When, at the time of my short trip in late June 2017, an overturned oil tanker spilt 25,000 litres of petrol over the highway that resulted in a massive

Figure 0.1 Study area in Pakistan.

explosion which instantly claimed 150 lives. Most of the people who died in the explosion were nearby villagers who had arrived at the scene to gather the "precious" leaking fuel in kitchen utensils. I witnessed the silent faces of young and old women and men, the people of the villages who seemed to question their state of ignorance and deprivation. In the following days, many national newspapers debated the poor living conditions of rural Seraiki people with an undertone of pity, yet they did not link the event to the existence of structural deprivation that has engulfed South Punjab since independence.

The governance of this region is mostly in the hands of civil servants, most of whom are from central Punjab, an educationally and economically stronger region. The presence of these Punjabi bureaucrats in the area, many locals argue, plays a major role in strengthening structural inequalities against the seventy per cent Seraiki-speaking populace. The politicians of the region, most of whom are Seraiki landlords or hold the status of spiritual leaders, remain self-absorbed and unconcerned with the welfare of their people. Over the years, they have done little to improve literacy, employment, infrastructure, and health care in the region (Khan 2004). In the past three decades, the feeling of deprivation has increased among South Punjabi people, leading some Seraiki activists to fiercely demand a separate province (Loyd 2020). Most "Seraiki nationalists" I spoke with over the years point to years of neglect and inequality by the provincial governments and stress the linguistic, geographical, and cultural distinction between Seraiki and Punjabi speaking groups to justify their demands for "Seraikistan" province (see also Rahman 2007, 175–178). Their demand has appeared at the federal level, however, despite promises made before each general election, the last three governments have opted to maintain strategic silence on the issue.

In current national discourse, *Seraikiness* is synonymous with *backwardness*, and impoverished conditions of the region are considered the root cause for the wide prevalence of polio and tuberculosis, child labour, and religious extremism in the region. For the urban, educated animal rights activists, as I discussed earlier, South Punjabi backwardness is the reason behind the attraction of rural men to engage roosters and dogs in brutal combat. To study such claims, I narrowed down my focus to the town of Kahror Pacca and multiple villages within the districts of Lodhran and Bahawalpur. These places (as shown in Figure 0.2) are the epicentres of pigeon flying, cock-fighting, and dogfighting, and could prove especially productive sites in gaining an insight into these practices. The centre of my fieldwork was the town of Kahror Pacca, whose square houses and narrow lanes turn the area into a dense, interactive place, despite the massive migration at the time of Partition in 1947 of the numerically dominant Hindus to India. Old Hindu temples and other historical buildings are still prominent landscape markers used by the locals to navigate the city. The city of Lodhran lies towards the west of Kahror Pacca, whereas Bahawalpur is towards the south.[4]

The Sutlej River divides Kahror Pacca and Lodhran from Bahawalpur. After crossing the sandbar of the river and beyond the terrains of Bahawalpur city lies the vast desert of Cholistan that shares the boundary with Indian Rajasthan. Rural Cholistani men generally wear white in summer (white turbans) and black in winter (black shawls), whereas the women don bright-coloured clothes and decorate themselves with arm-length bangles. North-west and north-east of the desert lie green patches of cultivated land and mango groves, irrigated by tube wells when the water level is low in the river and in its linked canals. Although I settled in one village, I travelled regularly to Kahror Pacca and to other villages on both sides of River Sutlej and some parts of Cholistan to meet the animal keepers and participate in their daily activities.

The extreme weather conditions of the area impact the human–animal relationship. The scorching midday sun of summer forces people to take shelter in their tiny adobe houses. Although the glare of the sun hurts the eyes, many pigeon flyers climb on to their rooftops and fly their pigeons to compete with opponents, shouting and whistling with gusto. However, cockfighters

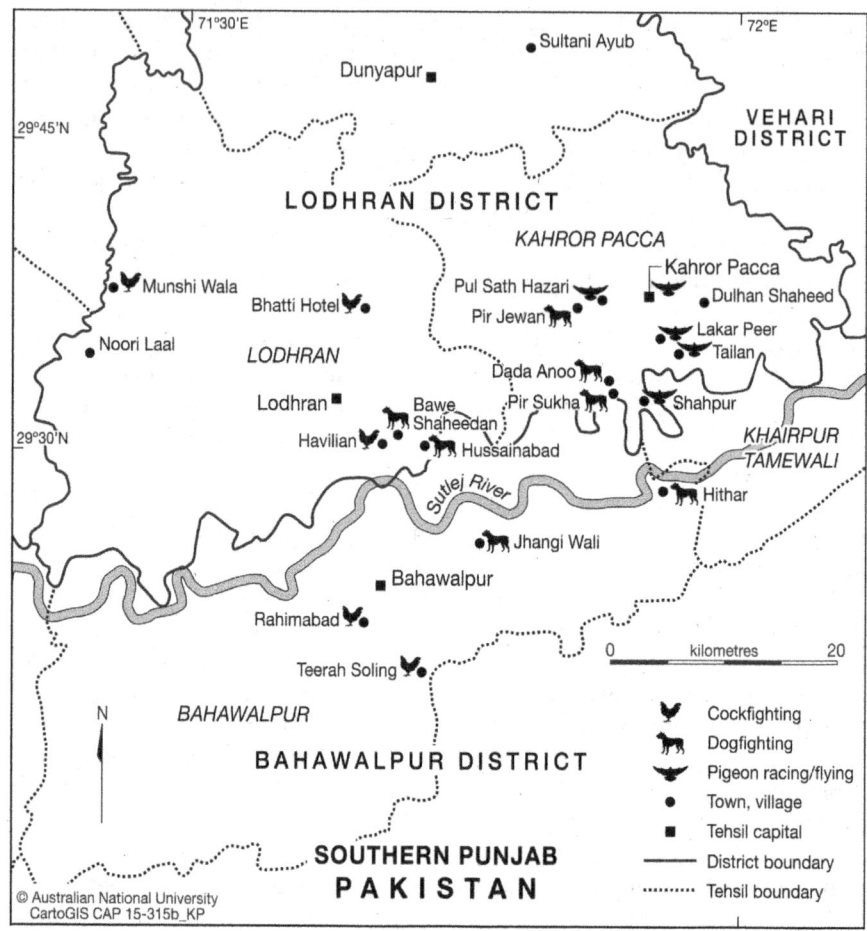

Figure 0.2 Pigeon flying, cockfighting, and dogfighting in South Punjab.

and dogfighters limit their activities in summer, training their animals only in the morning or in the late afternoon. In contrast, the village people maintain their relationship with cattle by keeping them in the shade of rosewood and banyan trees, lead them to bathe in the canals, and sleep close to them at night. Summer nights are quiet, with only the sound of nocturnal insects and barking village dogs piercing the silence. When the moon is high on summer nights, as the local saying goes, you can see an ant crawling on the ground.

Winter is dry and the months of December and January are cold and foggy, forcing the people to light fires in the mornings and evenings to keep themselves and their animals warm. During the day, animal keepers take part in cockfighting and dogfighting, along with numerous pigeon flying competitions. In dark wintery nights, many animal keepers huddle together around a crackling fire, smoke the hookah and share their stories. I was a regular participant in these impromptu gatherings, and listened carefully to how the men discussed their *shauq* and their aspirations to achieve masculine *izzat*. These meetings allowed me to think critically about the men's day to day relationship with their animals and their views about life, work, care, and passion.

Working with the *shauqeen*

During the twenty months of fieldwork, from August 2008 to April 2018, I witnessed rural South Punjabi life in full swing.[5] Everyone woke up early and the women left for the fields to pick cotton by wrapping huge cloth bags around their waists, while the men ploughed the fields with tractors to plant the next crop. Children generally began their mornings by walking lazily to school, and elderly men gathered at the local *chai* stall to be entertained by an old Bollywood action movie. At night, young boys wagered each other at the village pool table, while girls helped their mothers heat the clay oven in the courtyard of their houses to prepare dinner. As a "local," I presumed that my South Punjabi background, fluency in Seraiki language, and familiarity with South Punjabi "culture" would ease my research work with animal keepers. However, early on in my second immersive fieldwork in the fall of 2014, an event challenged this presumption.

As I started my fieldwork with dogfighters, Meena Darzi, an avid forty-year-old dogfighter and a tailor by profession, became my friend. He was the owner of a dog famous in the area, named Rambo, whose market value he claimed was 400,000 Rupees (about $2500).[6] "A man offered me his 1999 model Fiat 480 tractor (valued at 380,000 Rupees) in exchange for Rambo but I refused," he once told me. Meena planned to enter Rambo in the upcoming dogfight and invited me to accompany him. On the fighting day, Meena, his two friends, the thirty-three-inch high dog, and I, reached the arena in the car of one of his friends. Thousands of people thronged the arena in anticipation of the dogfight. Meena proudly introduced me to his friends as, "he is my *bhirā* (brother) here from Australia to write a book on Rambo." Meena, I later discovered, was an expert at quickly fostering fictive relationships by using kinship idioms. On this occasion, his pride in mentioning my "Australian" connection and my intention to write a book about *his* dog seemed to boost his status among peers.

The organising committee decided that Meena's dog would fight against Ashiq's dog that day. The two dogs fought ferociously until Ashiq's dog yelped and lost the combat. Meena delightedly picked up his dog on his shoulders and started celebrating the victory. Friends accompanied Meena in dancing; some showered ten-Rupee bills on him while others purchased rose garlands. In the ten minutes of celebration, everyone around him was basking in glory. I was intrigued by the whole "*tamāshā*" (spectacle) and was equally interested in the losing party. I approached Ashiq (who was the father of another friend of mine, Makhan Pehlwan) and asked him about his feelings after the defeat, his views on the competency of the referees, the possibility of foul play, the future of his injured dog, and his plans to avenge this defeat. Following my conversation with Ashiq, I made my way back to Meena's camp to find that he had already left the arena and was upset because of my "disgraceful act" (*be-gherti*). One of his friends explained that by not dancing with Meena after Rambo's victory, I had insulted him. Worse, I had gone over to speak to the opponent. "Meena called you brother and you dishonoured him," his friends concluded. I realised that I had made a dreadful mistake.

Two days later, I visited Meena and gave him a DVD that contained the video footage of Rambo's fight. The gift proved to be a good compensation for my erroneous behaviour. However, the incident opened my eyes to several issues. It highlighted the importance of acting carefully while conducting fieldwork and being mindful of how honour and shame play out in different contexts. It also indicated that fieldwork in an area familiar to the researcher is not always easy and straightforward. Most importantly, it showed that my status as an anthropologist did not necessarily afford me a licence to do what I wished; there were culturally acceptable behaviours I had to learn and abide by, and transgressions were not tolerated. Like the locals, I was also bound by the social norms, values, and customs of the area. In fact, because of my local background, there were additional expectations on my social conduct. The incident proved a turning point, influencing my conduct in the field and refashioning my methodological trajectory. It alerted me to the importance of exchange relations, of status and masculinity, and to the critical role of *izzat*.

In her essay, "How Native Is a 'Native' Anthropologist?" (1993) Kirin Narayan argues that anthropologists, both native and non-native, might be viewed in terms of shifting identifications. She suggests that researchers possess multiplex identities that are dominated by factors such as education, gender, sexual orientation, class, and race, and they outweigh our insider or outsider cultural identity (1993, 671–672). As I discussed above, the different strands of my identity (such as my education, city affiliation, Australian connection) rendered me an inauthentic insider and questioned my native status. Climbing on rooftops with pigeon flyers, sitting with cockfighters, or walking in the streets with dogfighters, I was reintroduced to South Punjabi norms and values. Like a humble neophyte, I learnt the values of keeping, training, and feeding the animals, and slowly recognised that my active interest in my interlocutors' lives proved a crucial methodological tool to explore the men's lives.

Because *shauq* is an important constituent of a person's life, regardless of gender, age, class, and occupation, learning about its articulation is something anyone can relate to in one way or another. It helped me understand and explore the wider cultural and social landscape of the animal keepers and their families. I learnt that their *shauq* is a passionate endeavour that provokes rich commentary on both their activities and wider cultural and social concerns, including their conceptualisation of *izzat*. This thread of *shauq* developed a sense of connection between us, and helped me collect life histories, conduct interviews, and paved the way for what anthropologists call participant observation.

In the following pages, I offer my experience of what it means to live and share a life with animals and their human keepers in rural Pakistan. I start with a survey of relevant historical data to explore the evolution, adaptation, and transformation of animal-related passions from pre-colonial India to modern day South Punjab (Chapter 1). I then move on to the description of my ethnographic encounters with pigeon flyers (Chapter 2), cockfighters (Chapter 3), and dogfighters (Chapter 4), and examine how humans and animals develop an inter-species relationship that shapes their lives, social roles, and understanding of others. In each of these chapters, I focus on the *shauq* of keeping, decorating, and breeding animals that are used to perform masculinity and secure prestige and *izzat* in the rural setting. However, a deep association with animals in an inter-species social dwelling raises important questions on the men's prioritisation of care, protection, and resources between their cherished animals and their close family members, and therefore in Chapter 5, I discuss women's experiences of living with animal keepers in a multi-species household. Finally, in Chapter 6, I examine the category of "genuine *shauq*" and various threats it faces in present times.

In closing, it is important to point out that I do not condone animal baiting or different forms of cruelty that animals may be subject to in rural Pakistan. An examination of animals' suffering in these practices is vital and perhaps the subject of an increasingly important scholarship that focuses on animal rights and animal liberation. However, I write this book to examine how human–animal lives unfold through a complex entanglement of care and violence in rural Pakistan. After 9/11, due to security concerns and the dearth of funding agencies willing to support fieldwork in Pakistan, there is a gap in the exploration of Pakistani culture (Ewing 2010). Most of the scholarship emerging from the country in recent years revolves around Islamic mysticism, fundamentalism, governmentality, and nationalism.[7] This book addresses that gap and takes the lens of human–animal relationships to provide a closer examination of everyday desires and passions of rural Pakistani men who shape their ordinary lives into an extraordinary space through their cherished animals, and structure a relationship with them to achieve honour (*izzat*) and reproduce masculinity (*mardāneat*) despite constant social and economic threats.

Notes

1 A limitation of this study is in my usage of the term "human–animal relationship," which includes men's relationship with their animals. Although I discuss women's relationship in Chapter 5, this data is not on women's relationship with animals but only their views on men's passion to keep animals.

2 For an extensive discussion on *shauq*, particularly in the context of South Asia, see Narayan and Kavesh (2019).
3 Broadly, the area known as "South Punjab" contains almost a dozen districts. However, in this book I take the districts of Lodhran and Bahawalpur as "South Punjab." I do this to emphasise the linguistic and cultural difference from central Punjab, and to relate animal activities to the poverty and underdevelopment of this area; see Figures 0.1 and 0.2.
4 Bahawalpur, a former princely state, got its name from Nawab Bahawal II in 1802.
5 This qualitative data was collected in South Punjab and through long-distance telephonic conversations included thirteen focus group discussions, ninety audio-recorded in-depth interviews, and eight life histories. My daily discussions with animal keepers were complemented by field notes and observations, and interpreted in the light of discussions with villagers and close relatives. The activities I describe in this book are restricted to the male domain, therefore, I spent the majority of my time with men, exploring their interpretations of *shauq*, their perception of human–animal relationships, and their ideas about masculinities and *izzat*. As a male researcher, interviewing women was difficult, however, to gauge the perspective of animal keepers' wives on the effects of male-centred human–animal relationships on the family, I took the assistance of two female Research Assistants.
6 Throughout this book, I use the dollar sign ($) to convert local currency into US dollars which at the time of writing this book was roughly 1/160, for example, $1 = 160 Rupees).
7 See Khan (2012), Ewing (1984), Hull (2012), Ahmed (2002), Frembgen (2011), Werbner (2003), Nelson (2011), Wolf (2006), Soares and Osella (2009), Philippon (2011), Jamal (2013).

1 Decolonising passions

From the sixteenth to the eighteenth centuries, pigeon flying, cockfighting, and dogfighting were all practised predominantly by the ruling classes in India. Over time, the elites (*nawabs*, princely chiefs, rajas, village landlords) refined these practices and their popularity grew among the common people who took on these pursuits as *shauq*. From their height to their transformation, these activities also demonstrated different types of masculinities. Masculinity, as Connell (1995) suggests, should not only be defined in relation to male domination of women but also male hegemony over other men. She emphasises the plurality of masculinities across a spectrum that includes hegemonic, subordinate, complicit, and marginalised masculinities, ranked hierarchically under the hegemonic model (1995, 76–80).[1] Connell's paradigm of the multiplicity of masculinities is particularly useful for analysing the transformation of recreational activities like pigeon flying, cockfighting, and dogfighting over time in the Indian subcontinent.

In the pre-colonial period, various animal practices (hunting, elephant fighting, cheetah fighting, and so on) were used by men in power to display dominant masculine traits such as courage, fearlessness, and strength. Among the Mughal kings, such practices were also a part of their everyday courtly recreation and leisure (O'Hanlon 1997, 12; Pandian 2001, 90). Thorstein Veblen in *The Theory of the Leisure Class* (1912) defines the concept of leisure as "non-productive consumption of time" (1912, 43). Leisure, in his view, is time that people spend in non-industrial (or labour) work. This, he says, serves two functions: (a) to assert the unworthiness of labour, and (b) as evidence of pecuniary ability to afford such a lifestyle. However, an important function of leisure activities that Veblen overlooks, later highlighted by Bourdieu (1984), is its ability to acquire symbolic gains, or its capacity to help a person achieve non-material goals, such as honour, prestige, status, and a position of distinction. For instance, the imperial practice of hunting tigers, usually carried out as a recreational pursuit from the sixteenth to the early twentieth centuries, served the Mughal kings (and later to the East India Company sportsmen) as a means of displaying their martial qualities as well as to assert the legitimacy of rule by offering protection to the Indian subjects (see Pandian 2001; cf. Orwell 1958). Similarly, as I will show in this chapter, the practices of pigeon flying, cockfighting, and dogfighting originally appeared as recreational pursuits, largely adopted by the Indian elites (sometime in collaboration with the British) to achieve symbolic gains and to reproduce traits of dominant masculinity.

Rosalind O'Hanlon (1997), a social historian, expands on the concept of masculinity to highlight the dimension of sociality. She views masculinity as "that aspect of a man's social being which is gendered: which defines him as a man and links him to other men, and conditions other aspects of his identity, such as class, occupation, race and ethnicity" (1997, 3). Masculinity for O'Hanlon is a social fact, linking one man to other men, and shaping his identity in a particular cultural setting. As I will show in the following chapters, the animal keepers of South Punjab strive to perform their masculinity by winning animal competitions and by accumulating status and *izzat* (honour). The masculinity of the animal keepers, therefore, is not only maintained by showing ascendency over other gendered identities but it also requires the recognition of other men (see O'Hanlon 1997, 3). However, an important feature of studying masculinities, as both O'Hanlon (1997, 3–4) and Connell (1995, 185) point out, is their evolution and transformation through history. This is one of the major concerns of the chapter: to explore the historical development of these three animal-related practices and explain how they canvas a plurality of masculinities that evolved since pre-colonial times, and in doing so to trace their transformation from a predominantly elite passion to a *shauq* of the rural people in contemporary Pakistan.

This chapter discusses the multiple forms of overlapping and contesting masculinities associated with all three animal activities in pre-colonial India. Pigeon flying was a passion of Indian kings, nawabs, and rajas, while the Indian elites and early colonial officials both practised cockfighting enthusiastically, dogfighting was introduced by the British to display their imperial masculine traits. All these practices were also effective social lubricants for Indian elites and the British and helped them develop social relationships. However, after the War of 1857,[2] relations between the rulers and the ruled were altered, and this had an effect on human–animal relationships in India.[3] These animal-related activities started losing their appeal to the British and consequently to the Indian elites in the late nineteenth century. This led to pigeon flying, cockfighting, and dogfighting entering the lives of the rural people and allowing them to compete for *izzat* among peers. The colonial legacy, however, still continues in post-colonial Pakistan in the form of the century-old *Prevention of Cruelty to Animals Act* which, even after some recent amendments, serves as a relic of colonialism.

Animal activities in pre-colonial India

Pigeon flying was practised among the people of India before the Mughals; however, the early Mughals elevated its status (Pearson 1984, 339). They exchanged birds to build political alliances and to develop and maintain friendships with the nearby states. The Mughal Emperor Muhammad Akbar (1542–1605) was particularly famous for keeping and flying pigeons, a passion he cultivated from his childhood. In the book *Ain-i-Akbari* (Akbar's Regulations), written by Akbar's court historian Abul Fazl,[4] an entire chapter is devoted to the king's passion for flying pigeons (Fazl 1873, 298–303). Akbar cherished the company of his birds, and referred to the practice as *'ishqbāzī* (love-play) rather than *kabūtar bāzī* (pigeon fancying)—the more common term used to describe the practice in modern times. The Arabic word *'ishq* means "to love passionately" or "be passionately in love"; however, it also means to "interjoin closely," and to "connect" (Wehr 1979, 719). Because of the word's lexical richness, it is used by many Persian and Indian Sufis to hint at their union with God (Lumbard 2007, 373). Akbar, too, took his *'ishqbāzī* to mystical heights, as an appreciation of the wonders of the Divine Creator, as Fazl notes:

> The amusement which His Majesty derives from the tumbling and flying of the pigeons reminds of the ecstacy [sic] and transport of enthusiastic dervishes: he praises God for the wonders of creation. It is therefore from higher motives that he pays so much attention to this amusement.
>
> (1873, 298)

The number of pigeons at Akbar's court was staggering. It is estimated that he had more than 20,000 pigeons and almost 800 kilos of grain required to feed them each day (1873, 302). Many enthusiasts still refer to Akbar's practice of feeding pigeons with seven types of grain to keep the bird healthy and active. Moreover, he employed many servants whose only job was to care for pigeons, and they were paid a salary equivalent to trained soldiers (1873, 303). Such allowances demonstrate the importance of the bird, which was not only kept as a form of entertainment but also for strategic purposes.

Throughout his reign, Akbar took personal interest in the birds and used selective breeding. He gave them personalised names and trained them to fly at night. He is also said to have been the first bird trainer to fly a cluster of more than 100 pigeons at once. In addition, Akbar introduced new methods for evaluating pigeons (for instance, the colour of the bird's eye, its claws, and sides of its beak), and developed the typology of pigeons based on their thirty colours and fifteen patterns (1873, 301). Fazl describes Akbar's 500 selected pigeons, known for performing remarkable manoeuvres such as doing seventy tumbles in the air. However, these feats were not only designed to amuse the king because Akbar deployed the bird strategically to facilitate matters of state. According to Fazl, pigeons were commonly used to impress his subjects and foreign visitors. The king would astonish merchants and diplomats coming from Iran and Turan with the remarkable displays of skill of his pigeons (1873, 289–299).

As a court historian, Fazl always found crucial functional justification for the king's pursuits. For instance, while discussing the game of *chaugān* (Indian form of polo), Fazl suggests that it was not merely a form of amusement, but that the king saw it as a training practice for the Mughal cavalry (1873, 297). Again, in his discussion of the king's hunting expeditions, Fazl claims that for Akbar it extended well beyond recreation, as it allowed him to increase his knowledge about the state, military, and Indian subjects (1873, 282). This was equally true of pigeon keeping, which Fazl claimed was an important affair:

> This occupation affords the ordinary run of people a dull kind of amusement; but His Majesty, in his wisdom, makes it a study. He even uses the occupation as a way of reducing unsettled, worldly-minded men to obedience, and avails himself of it as a means productive of harmony and friendship.
>
> (1873, 298)

Keeping an extremely large number of pigeons at court not only displayed Akbar's power and wealth, it also depicted the king's ability to control, tame, and domesticate the wild birds through his care and attention. This symbolic care of the birds, Fazl's description indicates, was analogous to the king's care of his subjects and visitors who, like pigeons, were carefully nurtured.

Two centuries after Akbar, pigeon flying continues to be associated with the ruling elites of India. Mughal miniaturists illustrate the bird, and different literary works affectionately depict pigeons. One such artwork, *Kabūtar-nāmah* (The Book of the Pigeon), appear in the mid-eighteenth century, written by a late Mughal author, Vālih Mūsavī, containing colourful illustrations, evocative poetry, and vivid prose concerning pigeons (Sims-Williams 2013).[5] After its metrical introduction, the prose in the second section of the book details breeding, training, and flying techniques, and elaborates on the various colours and patterns, characteristics, and descriptions of the bird.[6]

Pigeon flying remained popular among the Indian kings, nawabs, rajas, and elites until the mid-nineteenth century. The last Mughal king, Bahadur Shah (1775–1862), was also said to have been fond of pigeons.[7] Legend has it that when the king went out in procession, almost 200 pigeons flew overhead to provide shade from the scorching summer sun of Delhi (Sharar 1975, 127). The rulers of Oudh, from Shujaud Daula (r. 1754–1775) to Nasirud din Haidar (r. 1827–1837) were all pigeon enthusiasts, and the passion reached its peak during the reign of the last nawab of Lucknow, Wajid Ali Shah (r. 1847–1856). There are important similarities worth noticing in the passion of pigeon keeping between Nawab Wajid Ali Shah and Emperor Akbar. Both kept more than 20,000 pigeons at their courts and employed around 300 caretakers for the birds. Like Akbar, Wajid Ali Shah had a large and impressive menagerie (Sharar 1975, 128).[8] However, unlike Akbar whose enthusiasm for pigeons has been described as a "study," for achieving "higher motives," and for reducing "worldly-minded men to obedience"; Wajid Ali

Shah's *shauq* is usually depicted as an extravagance and a wasteful practice. Colonial accounts note that he once paid 25,000 Rupees for a silk-winged pigeon and spent 2000 Rupees on producing a pigeon with one black and one white wing (see Oldenburg 1984, 15–16). The shifting status of pigeon keeping—from being a demonstrative of higher status to a mark of wastefulness and extravagance—was in line with many other traditional pursuits which the British rulers dismissed as frivolous and inefficient. This is poignantly depicted in Satyajit Ray's critically acclaimed film "The Chess Player" where the British Resident of Oudh is shown to be extremely impatient and dismissive of the local nawab's love of poetry, dancing, pigeon flying, and kite flying.

By the first half of the nineteenth century, the ideals of imperial rule and notions of masculinity circulated within a wider discourse that championed ideas of productivity, order, and authority. The "wasteful use" of resources, energies, and time by the local elites on activities such as pigeon flying was seen as yet another indication of their indolent and effeminate character. The idea of waste and more specifically of moral waste, as geographer Vinay Gidwani (1992, 44) has described, served as a metaphor for the idle behaviour of the Indians and to some degree became a political tool for legitimating colonial rule in India. Nevertheless, keeping and flying pigeons was practised in some parts of England from the sixteenth century onwards. For example, a small community of the silk weavers of Spitalfields kept and bred pigeons for many generations.[9] However, the passion of this artisan community was never considered a gentleman's pursuit. Charles Darwin, who used to visit the place to discuss pigeon breeding, described the locals as "little men" and "odd specimens of the Human species" (Feeley-Harnik 2007, 161–162). As the practice was out of step with Georgian gentlemanly traits, it remained marginal, far removed from the newly assumed authority that the British Raj sought to establish in its colonies.

In short, unlike the British colonial officials who considered it an indolent pursuit, in India pigeon flying was a culturally meaningful activity, associated with elite masculinity that demonstrated high culture, authority, and financial display. Contemporary Indian social theorist and political psychologist, Ashis Nandy (1983, 10), suggested that the conception of masculinity in pre-colonial India was quite different from the European imagination of gender. What the Europeans considered "effeminate" behaviour, such as the self-denying ascetic Brahman, a poet prince, a nawab listening to court music, or a king flying pigeons or kites, was not strictly identified as such in India. Some forms of creativity like poetry, song, and dance did not involve the display of aggressive behavioural traits, yet the Indians considered them vital for the cultivation of an ideal elite male. Similarly, pigeon keeping presented a notion of masculinity that, without the outright pursuit of power and domination, required years of study, care, and attention in mastering the craft. This was not comprehensible to the British who usually associated it with inefficiency, effeminacy, and idleness.

Cockfighting, however, was an activity that the British enjoyed and practise along with the Indian rajas and nawabs, and even wagered enthusiastically on fighting roosters. Unlike pigeon flying, cockfighting—for both Mughals and the British—was a portrayal of courage, bravery, and strength. The gamecock, like a soldier in the battlefield, was expected to fight with strength and courage to the last and not to surrender or flee from the battle. In fact, cockfights were purposely shown to soldiers before an armed engagement to raise their spirits and inspire them to fight like a cock, to their last breath.[10]

In many societies, roosters were a vital part of mythology and magic, considered a symbol of beauty and power, and usually associated with masculine virility (Donlon 1996, 186; Smith and Daniel 1975, 53). The birds were highly praised for their aggressive spirit and unyielding bravery and to witness such masculine traits, the first cockfight was organised almost 3000 years ago in South East Asia (Smith and Daniel 1975, 69; Donlon 1996, 187; Beebe 1921, 206). This makes cockfighting one of the most ancient forms of animal sport (Dundes 1994, vii). As the activity spread to other parts of the world, it reached Europe where gamecocks offered a model of masculine behaviour, displaying values of loyalty, courage, and diligence. Many British monarchs including Henry VIII, James I, Charles II, William III, and George IV were avid cockfighters in their times (Smith and Daniel 1975, 87).

In India, the activity was practised both by the common people and by the rulers (Harris 1994, 10). It was a part of Mughal court life, and a *shauq* of nawabs and rajas of different states. In the mid-eighteenth century, when the passion of cockfighting was at its peak in India, Vālih Mūsavī, the author of *Kabūtar-nāmah*, also wrote *Murgh-nāmah* (The Book of the Chicken) [Storey 1977, 410].[11] By the late eighteenth century, when the British had established their rule in India, British traders and officers fought their "English Game Fowls" against Mughals' "*Aṣīl*" breed.[12] Abdul Halim Sharar, an Indian historian who lived in Lucknow, observes that in Oudh, both the Indian rulers and the British participated in weekly cockfighting contests. "General Martin was an expert at cockfighting and Navab Sadat Ali Khan used to bet his cocks against those of the General," notes Sharar adding that in Matiya Burj cockfights near Calcutta—the centre of British rule at that time—British officers used to bring their birds to fight (1975, 123). Cockfighting was thus an important site for the negotiation of power for local elites and their British counterparts.

Cockfighting provided an opportunity to the Indian elites and the British to socialise, integrate, and canvas their masculine traits. However, the more virile form of establishing masculinities was considered to be the practice of fighting canines. Unlike cockfighting, dogfighting was European in origin, and represented popular masculine qualities. For instance, in Rome's Coliseum, the practice of baiting dogs, generally against bears, bulls, or badgers, demonstrated the fierce and aggressive spirit of canines (Gibson 2005, 4; Kalof and Taylor 2007, 321–322). In medieval England, particularly from the twelfth century onwards, dogs were considered a symbol of determination, power, and bravery—characteristics associated with English masculine prowess (Kalof and Taylor 2007, 322). Because of these attributes, the large and ferocious mastiff was usually used as a war-dog (Brownstein 1969, 243).

In the sixteenth and seventeenth centuries, baiting sports in England were more popular than going to theatre, and were considered to be a display of hyper-masculinity (Brownstein 1969, 237). These activities not only showed the English mastiff's ascendency over other animals such as bulls and bears, but also served an instrumental objective (Evans et al. 1998, 827). The bull, for example, was baited against dogs with an objective of tenderising the meat before slaughter, while the bear was selected due to its resemblance to humans in form and shape (Evans and Forsyth 1997, 61; Kalof and Taylor 2007, 322). The canine's triumph over the large bear displayed the dog's ability to outmanoeuvre a more strong and powerful opponent. There were also some instances where specially bred and trained dogs were staged to kill rats in rat-pits, making *ratting* a "gambling-sporting entertainment for the urban poor" (Anderson 2010, 30). The activity also offered "new ways of relating to rats," and provided opportunities for employment to many rat-catchers who by supplying rats for the contest turned the animal into "a desirable commodity" (Pemberton 2014, 528–530). However, the functional reasoning for different forms of baiting was secondary to their entertainment value. In all these dog baiting activities (whether against bull, bear, or rats), the crowd was captivated by the dog's fierceness and appreciated his virile qualities (Brownstein 1969, 243). Dog baiting in that period was not only popular amongst the lower or middle classes, it was also enjoyed by the royalty. Scholars note that King Henry VIII and Queen Elizabeth I enjoyed such pursuits, and presented mastiffs as royal gifts to other monarchs (Kalof and Taylor 2007, 322; Evans and Forsyth 1997, 61; Brownstein 1969, 243).[13] Such gifts of large and furious dogs symbolised the power of the giver—the powerful and authoritative Queen or King—and helped in establishing and reinforcing their royal reputation and standing.

In India, Mughal Emperor Akbar and his descendent Jahangir (1569–1627) received dogs from European traders as gifts (Fazl 1873, 290). Akbar greatly admired his canines and used them for racing and hunting. Although some dogs were used to attack other animals during the hunt,[14] we find no record of dogfighting or bear baiting in the early Mughal era.[15] The British most likely introduced dogfighting and bear baiting during the Company Raj. The traces of this legacy persist, as during my fieldwork I found English bull terrier (and their outcrosses, such as the Gull Terrier) being used in animal baiting in Pakistan.

During my ethnographic fieldwork in South Punjab in 2014–2015, a number of respondents reported that it was the British who had introduced specially trained bull terriers to fight against bears. This view is substantiated by a publication of the World Society for the Protection of Animals which states that the perpetuation of bear baiting in India was a British tactic to build political capital and relations with Indian rulers, and to "establish their important status within the rural community" (Joseph 1997, 4; see also Bio-resource Research Centre 2010). Other than developing social relations, the baiting activities were a platform that enabled the British to perform imperial masculinity publically. The fighting dog, like an English soldier, was considered brave, strong, disciplined, and ready to fight at all times.[16]

In the eighteenth century, most baiting activities (including bull baiting and bear baiting) were on the decline. Dogfighting replaced these as the popular form of amusement, most likely because of its more fierce and competitive nature (Kalof and Taylor 2007, 323; Evans and Forsyth 1997, 62). In addition to participating in dogfights, some people among the elite began breeding fighting dogs commercially and were the chief lobbyists against vocal demands by animal welfare groups to ban the practice (Evans and Forsyth 1997, 62).[17] However, by the late nineteenth century, dogfighting and other baiting activities gradually lost their elite patronage in England,[18] and were outlawed with the Prevention of Cruelty to Animals Act of 1835 for multiple reasons (Coleman 2009, 88; Kalof and Taylor 2007, 323).[19] Firstly, the Church of England opposed such activities because they compromised church attendance on Sundays (Evans and Forsyth 1997, 62); secondly, the Victorian ethos—self-righteousness, morality, and sympathy—were employed to argue against these "barbaric activities"; and, thirdly, the "English gentleman" was changing with time and engaging in more competitive sporting activities that involved the wilderness and adventure (hunting), stamina and athleticism (cricket and hockey), and danger and courage (chasing man-eating tigers).

Animal activities in late colonial India

From the eighteenth century onward, European moral philosophers debated over non-human suffering and ethical use of animals. At that time, the streets of London were crowded with animals; some were exhausted livestock used for labour or destined for slaughter, others were baited for human amusement (Ritvo 1994, 106). The main purpose of these debates was to sensitise people to different forms of animal cruelty. One of the pioneers of these debates was the English utilitarian philosopher Jeremy Bentham (1748–1832), who advocated for animal welfare by famously arguing: "The question is not, Can they [animals] reason? nor, Can they talk? but, Can they suffer?" (See Benthall 2007, 2). These debates later developed into the animal welfare movement, and eventually led to the establishment of "The Society for the Prevention of Cruelty to Animals" in England in 1824. Almost sixteen years later in 1840, the society received royal patronage and changed its name to "The *Royal* Society for the Prevention of Cruelty to Animals" under the aegis of Queen Victoria, thus attracting substantial political clout to advocate for animal welfare (Ritvo 1994, 108; see also Maehle 1994).

Inspired by the ideals of the Enlightenment, the animal welfare movement was significant in shaping the English middle class's attitudes towards animals. Although the movement discouraged animal baiting and beating, it did not attend to all types of animal cruelty. The animal rights movement, which gathered pace in the latter part of the twentieth century, proved broader and holistic in its demands for non-humans. Heavily influenced by Kantian ethics, the animal rights groups reasoned that animals (like humans) must be treated as ends in themselves, and not as a means (Sunstein 2004, 5; Francione 2004, 16). Unlike the animal welfare movement that preceded it, animal rights aimed to stop all types of human abuse of non-human animals, whether in the domains of agriculture (farming and food), clothing (wool and silk), entertainment (circus, zoo, or baiting), experiments (in laboratories), or hunting (including fishing).

The most significant contribution of animal welfare groups of the eighteenth and the early nineteenth century is their role in developing laws and halting animal baiting activities in England. These laws were subsequently introduced to the colonies. In India, *The Prevention of Cruelty to Animals Act* (1890) officially outlawed different forms of animal cruelty, including animal baiting for amusement. Although this law clearly reflected the efforts of animal welfare groups, it was also supported by the ideals of Victorian moralities that were increasingly shaping British rule in India in the late nineteenth century. Queen Victoria's reign was characterised by an ideological and moral shift in English society, as it introduced ethical considerations that sought the extension of sympathy to all beings, self-improvement, productiveness, the controlling of "natural" impulse, morality, duty, and the expression of gentlemanly character. It is no surprise that under such a new value system, the impulsive activities of baiting animals lost favour, replaced by more composed activities afforded to the ruling elites, including cricket, going to theatres, and visiting museums and exhibitions (cf. Bailey 2007, Vorspan 2000).

Among these activities, cricket was especially important. In England, cricket was an expression of the Victorian elite values; a quintessential male sport practised to cultivate masculine behaviour through sportsmanship, fair play, unquestioned loyalty, control, and subordination of one's personal sentiment (Appadurai 1996, 91–92). While writing about the value of cricket in the Victorian society, Keith Sandiford (1983) notes:

> Cricket was much more than just another game to the Victorians. They glorified it, indeed, as a perfect system of ethics and morals which embodied all that was most noble in the Anglo-Saxon character. They prized it as a national symbol, perhaps because—so far as they could tell—it was an exclusively English creation unsullied by oriental or European influences. In an extremely xenophobic age, the Victorians came to regard cricket as further proof of their cultural supremacy.

The British took cricket as a vehicle to impart the Victorian ideals of "fitness," "stamina," "manliness," and "vigour" into what they saw as the "effeminate," "lazy," "enervated," and "effete" Indians (Appadurai 1996, 93). The indigenisation of cricket, anthropologist Arjun Appadurai observes, enabled the colonial authorities to discipline the natives, and linked Indian princes with the English aristocracy in novel ways (1996, 94). To maintain their social linkages with the British, the princely rulers enthusiastically promoted this new sport (Khan and Khan 2013, 2–3; Majumdar 2006, 887–909; 2003, 171), slowly ceasing their patronage for traditional pursuits including pigeon flying, kite flying, cockfighting, and wrestling.

Cricket also promoted notions of muscular Christianity. Clergymen were convinced that a healthy mind and body were essential for cultivating spiritual qualities and thus promoted the ideals of discipline and exercise among Victorian youth. For them "Godliness and manliness" or "spiritual perfection and physical power" were intertwined, as they strongly believed that "a feeble body could not support a powerful brain" (Sandiford 1983, 305). Since many clergymen also served as headmasters of English public schools, the educational institutions implemented the ideals of athletic masculinity to produce civic leaders (Sandiford 1983, 306). The game of cricket, along with other British sports, thus provided a practical arena for the implementation of the Victorian ethos (control of impulses) and expression of Anglo-Protestant masculinity (disciplining the body).

In the latter half of the nineteenth century, physical strength, muscular power, and aggression were not the only traits that were valued as the dominant form of masculinity in India. During the Raj, a new colonial masculinity was formed in complex interrelations between residual and emergent traits. The Marxist literary theorist, Raymond Williams (1977, 122–123), describes such relations in his discussion of "Dominant, Residual, and Emergent," where he maintains that the "residual" or elements of any culture's past interact with the "emergent" and new meanings, values, practices, and relationships are formed within a culture. The dominant constantly influences the residual and the emergent. By employing these insights, we might say that in nineteenth century India, the residual traits of masculinity were courage, stamina, power, aggression, and so on; and the emergent traits included the new gentlemanly masculine traits such as discipline, loyalty, self-control, and governance. The amalgamation of residual and emergent traits was a new type of colonial masculinity shaped by Christian ideals and thus assumed a superior position in the gender order of late nineteenth century India.

The ascendency of this new form of masculinity, along with the influence of the Victorian ethos, led the Indian elites to withdraw from practising pigeon flying, cockfighting, and dogfighting. Pigeon flying was now considered a wasteful activity, and cockfighting and dogfighting were viewed as "immoral" practices under the dominant Anglo-Protestant value system. At the end of the nineteenth century, such considerations paved the way for the British Government of India to outlaw different forms of animal cruelty, including cockfighting and dogfighting, by adopting the 1835 version of the

Prevention of Cruelty to Animals Act with some changes. In 1890, this *Act* officially banned all forms of animal baiting, as the section 6(c) read:

> If any person—(a) incites any animal to fight, or (b) baits any animal, or (c) aids or abets any such incitement or baiting, he shall be punished with fine which may extend to fifty rupees. Exception: It shall not be an offence under this section to incite animals to fight if such fighting is not likely to cause injury or suffering to animals and all reasonable pre-cautions are taken to prevent injury or suffering from being so caused.

Though the section only stated a fifty-Rupee fine, section 3(a) mentioned one month's imprisonment for the one who:

> Overdrives, beats, or otherwise treats any animal so as to subject it to unnecessary pain or suffering.

However, it was the ruling class that determined what constituted pain or suffering. That is, while popular activities such as cockfighting and dog-fighting were banned; game shooting, pig sticking, and tiger killing continued unchecked. These were not described as "immoral" pursuits but as refined gentlemanly recreations. The case of tiger killing is important to note here. For the Mughals, anthropologist Anand Pandian observes, killing tigers helped in displaying their ferocity to win over the obedience of unruly local chiefs (2001, 90). For the British, however, tiger killing rendered them as the masters of the wild and "caring and responsible sovereigns" of native Indian subjects (2001, 80). Even in the twentieth century, "refined" pursuits like tiger killing and other hunting activities were practised by the British officials and military officers through the provision of special hunting licences (Pandian 2001, 83; Hussain 2010, 120). Thus, the law that banned cockfighting and dogfighting did little for the protection of other species.

Concluding remarks: animal activities in present-day Pakistan

Above, I have provided a brief history of pigeon flying, cockfighting, and dogfighting to demonstrate their adaptation and development from elite-based activities to enthusiastic pursuits adopted by the general public of the subcontinent. I have showed that these animal activities were not merely court recreation or indolent leisure, but served as important markers of power, instrumental in securing social status, forging diplomatic links, and in articu-lating different forms of masculinity.

Pigeon flying was adopted and refined by early Mughals like Akbar, who took it to new heights to impress foreign dignitaries with his artful care and taming of the wild. Cockfighting was practised by both the Indian elites and the British as a political contestation of masculinity whereby each party sought to display its supremacy over the other. Dogfighting was a British practice, introduced to India as part of a wider programme to develop relationships with Indian elites and to demonstrate British authority through a spectacle where imperial courage, bravery, and strength were on display. From the nineteenth century onward, as the British began to promote Victorian values through sports like cricket, and with the emergence of Anglo-Protestant masculinity alongside the formation of the *Animal Protection Act* in India, there was a marked decline in pigeon flying, cockfighting, and dogfighting.

After the partition of India in 1947, the newly established countries adopted the 1890 version of the Act. The Indian Government revised the Act in 1960 (Srinivasan 2013, 108), however, in Pakistan it remained in its original form for more than a century. Only recently in January 2018, the Pakistan Government amended the amount of penalty for inciting animals to fight from a fifty-Rupee fine to a 300,000-Rupee fine (about $1900). However, except for altering the amount of penalty, this amendment did not change the overall structure of the *Act*, and animals are still not recognised as sentient beings with intrinsic values (Shah 2018). This suggests that although some politicians have responded to the increasing demands of animal welfare groups, the *Act* continues to serve as a relic of colonialism in post-colonial Pakistan.

In practice, there is minimal enforcement of this *Act* by the police. For instance, in the Punjab province, there are two government departments concerned with animals: "The Livestock and Dairy Development" department (known as the Livestock Department), and the "Forestry, Wildlife, and Fisheries" department (known as the Wildlife Department). Pigeons, chickens, and dogs come under the Livestock Department. However, due to insufficient funds and lack of political will, the Livestock Department is occupied with veterinary affairs and hardly pursues those engaged in cockfighting or dogfighting. It has minimal human resources available and possesses no power of enforcement. Conversely, the Wildlife Department may have the power to stop animal abuse, but its jurisdiction is limited to wild species. The key contribution of the Wildlife Department is when it, in coordination with Bioresource Research Centre (BRC), succeeded in halting bear baiting in rural areas. The underlying motivation was clearly to save the Asiatic black bear, a "vulnerable" species according to IUCN (2015) that was extensively poached and used in circuses, for performing tricks in streets by bear trainers, and most importantly, involved in baiting by village landlords who would unleash bull terriers on a tethered bear to reproduce their symbolic authority in the village environment.

To this day, pigeon flying, and fighting roosters and dogs (animals, who unfortunately, are not "wild" and do not appear on IUCN's red list, thus remain out of jurisdiction of the Wildlife Department and law enforcing authorities in Pakistan), remain popular in rural parts of Pakistan. The case of dogfighting, an activity that takes place in a public setting, is important to note. During my fieldwork, I attended almost eighty dogfights and the police raided none. On several occasions I noticed police officers (in uniform) merged with the large crowd, enthusiastically watching dogs fight. However, cockfighting was subject to occasional police raids, with the purpose of stopping gambling on fighting roosters but not cockfighting per se.

Gambling (*jowā*) is a socially immoral practice in Pakistan, and there is a religious prohibition to it. For example, the Qur'ān (2, 219; 5, 91) and many sayings from the Prophet consider gambling a grave offence, something that turns people into addicts and promotes ill feelings of jealousy and hatred among them. Hence, the Islamic Republic of Pakistan has outlawed all forms of gambling, including any type of activity in public or private settings. Such violations incur the minimum penalty of one-year imprisonment and/or a 5000-Rupee fine (about $30). It is also true that many Islamic accounts discourage and even prohibit animal baiting, however, it seems that the proponents of Islamic Sharia, particularly those who in late 1970s went out on a mission to Islamise the country and succeeded in banning gambling, were more concerned with human moral elevation than with animal suffering.

In current Pakistan, as the following chapters show, many rural people engage in pigeon flying, cockfighting, and dogfighting to display dedication, skill, courage, and bravery and try to reproduce the manliness of pre-colonial kings and rajas, and colonial administrators. With great gusto, the men take such activities as their *shauq*, and enthusiastically fly and fight their animals to achieve *izzat* (honour, prestige, and status) and to display various traits of rural South Punjabi *mardāneat* (masculinity). They cherish their practices and forget their self in these pleasures, thus temporarily escaping from the tribulations of everyday life. This is particularly true among pigeon flyers, many of whom keep, train, and fly pigeons on their rooftops to fulfil their personal passion, and develop a more-than-human relationship that carry crucial meanings in their social lives.

Notes

1 The particular traits of dominant masculinities change over time and across culture, see for example, the cases of the effeminate Jew of the Diaspora to the Zionist new man (Hirsch 2015, 305), martial and literary masculinity in China (Kipnis 2011, 120), or the changing forms of Irish and Serbian masculinities (Banerjee 2005, 9–10).

2 Also known as the Great Rebellion, or in colonial terms as "the Indian Mutiny," or "Sepoy Mutiny."

3 The War of 1857 officially marked the commencement of the British Raj from the Company Raj. During the British Raj, British interaction with Indians reduced greatly and scientific racism (Risley 1915) and Orientalist discourse (Said 1978) gained ground, driving a wedge between the ruling foreigners and native Indians. Meanwhile, gentlemanly sporting societies emerged that discouraged the participation of local people (e.g., Sinha's [1995, 86] discussion of Bengalis' "effeminacy"; they were deemed unfit for shooting tigers).

4 *Ain-i-Akbri* (Akbar's Regulations) is the third volume of the book series "*Akbarnamah*" (The book of Akbar). Other than containing official administrative reports, the book details the life and activities of Akbar, who ruled from 1556 until his death in 1605. Abul Fazl (1551–1602), the writer of *Ain-i-Akbari*, was Akbar's court historian, advisor, and chief propagandist (Pandian 2001, 91).

5   James Lyell in his *Fancy Pigeons* (1887, 412) translated the term *Kabūtar-nāmah* as "History of Pigeons." In the conclusion of his book, he added a few pages of Sir Walter Elliot's translation of *Kabūtar-nāmah*. Elliot collected the treatise from Madras and sent it to Charles Darwin to aid his research (1887, 404). Darwin in his *The Variation of Animals and Plants under Domestication* (2010, 41) regarded *Kabūtar-nāmah* as a "curious treatise."

6   We learn more about this book from Charles Storey's bio-bibliographical survey of Persian literature, where he mentions that Mūsavī was born in Khurāsān and migrated to Hyderabad, and later to Arcot (Storey 1977, 410–411). It was most likely during Mūsavī's time in Hyderabad that he wrote *Kabūtar-nāmah*. Dalrymple (2002, 264) mentions pigeon keeping as a popular pastime among the elites of Hyderabad at the end of the eighteenth century, and describes pigeon cotes as an "essential part of the cultivated enjoyment of a gentleman's pleasure garden."

7   Because of his interest in different *shauq* of the mid-nineteenth century (such as poetry, pigeon flying, kite flying, music, painting), Bahadur Shah chose "Zafar" and "Shauq Rang" as his pen names. He used Zafar (meaning victory) for his poetry and signed his music as Shauq Rang (an enthusiast of colours). See Welch (1985, 429) and Pritchett (1994, 5).

8   Abdul Halim Sharar's book was originally published in Urdu with the title *"Guzishta Lucknow"* (c. 1920), and later translated into English by E. S. Harcourt and Fakhir Hussain in 1975 as "Lucknow: The Last Phase of an Oriental Culture." In the Urdu version, Sharar used the word *shauq* on numerous occasions (such as when discussing animal fights, bird fights, kite flying, and pigeon flying). In the English translation, we do not find italicised "*shauq*," and the word is mostly translated as predilection, desire (1975, 116), interest and founding (1975, 123), and pursuit (1975, 131).

9   Charles Dickens also mentioned the traditional practice of breeding, keeping, and fancying pigeons in Spitalfields in his *Household Words* (1851).

10 This tradition of staging cockfights to inspire soldiers in the army goes back to the Athenian general, Themistocles, who emphasised the glory and fighting will of gamecocks in a speech to his soldiers before defeating the Persians. Later, he made cockfighting a national sport of Greece (see Smith and Daniel 1975, 70–71).

11 The treatise is a poetic work on cockfighting and contains information on the breeds and characteristics of roosters, the birds' dietary and training regimens, and the medication and fighting practices of cocks. Mūsavī's Persian version of *Murgh-nāmah* was translated by Nawab Yar Muhammad Khan into Urdu in 1883. From this Urdu translation, Lieutenant-Colonel D. C. Phillott translated *Murgh-nāmah* into English for the Asiatic Society of Bengal. American Ornithologist William Beebe in his *"A Monograph of the Pheasants"* included some parts of Phillott's translation (1921, 199–203).

12 An early administrator of East India Company, Robert Clive (1725–1774), along with other members of the British aristocracy, is believed to have cherished cockfights (Smith and Daniel 1975, 87–92).

13 In Liza Picard's (2003, 125–126) *Dr. Johnson's London: Everyday Life in London 1740–1770*, we also find some discussion of animal baiting, particularly cock-fighting, bull baiting, bear baiting, and badger baiting.

14 In Mughal amphitheatres, there were regular displays of animal fighting. These activities were staged to demonstrate the control and aggression of the king to the audience (Pearson 1984, 340). After the mid-nineteenth century, when royal

courts ceased to exist, these activities disappeared. Although we can find evidence of elephant fighting, camel fighting, ram fighting, cheetah fighting, and others in Mughal India, I have not come across any reference to dogfighting or bear baiting.

15 Personal correspondence with Ursula Sims-Williams, Lead Curator of Persian collections at the British Library, UK (20 July 2016).

16 Bull baiting did not take hold in India because bulls held religious and utilitarian (farming) significance for both Hindus and Muslims. Dogfighting and bear baiting, however, still survive among rural people in Pakistan.

17 For instance, when in 1800, the House of Commons presented a bill in the British Parliament for abolishing bull baiting, it was defeated on the grounds that such entertainment inspired courage and produced an elevation of mind among men (Ritvo 1994, 106).

18 According to the *Encyclopaedia of Traditional British Rural Sports,* dogfighting was practised in some parts of England until the end of the nineteenth century (Lile 2005, 102).

19 The *Ill-treatment of Cattle Act* was the first law of this kind that prevented cruel and improper treatment to animals (cattle), and was passed by both houses of the English parliament in June 1822 (Radford 2001, 39). Later in 1835, *the Animal Protection Act* (focusing on all animals rather than on cattle) was passed in England.

2 Living with pigeons
Rooftop intimacies

In late May 2015, news about a "spy pigeon" caught the attention of many pigeon enthusiasts in South Punjab. The national newspapers and some international news media reported that a pigeon was arrested by the Indian police for landing in "enemy" territory near the Pakistan border (BBC 2015). On suspicion of carrying a secret message from Pakistani intelligence agencies, the Indian police X-rayed the pigeon and kept it under close observation. Some of my pigeon flying friends ridiculed the news, while others took it seriously and debated the crucial role pigeons have played in battles throughout history, particularly in WWI.[1] A third group had a different take altogether and held that pigeons did not recognise man-made boundaries of India or Pakistan; they did not care for either Muslims or Hindus and were birds of peace who lived in both the "masjid" (mosque) and the "mandir" (temple), fed and loved by all. Among them was Rafik, who was adamant that pigeons do not discriminate between religions, nations, or people, and that this was just an excuse to perpetuate tensions between the two neighbouring countries.

In 1947, Rafik's parents, like many other Muslims, migrated from India and settled in the town of Kahror Pacca. "At the time of Partition, when people were bringing their most cherished belongings, my father carried two pigeons with him," Rafik proudly added that the *shauq* of pigeon keeping ran in his blood for many generations. He was a thin man of a short-height, not more than 45 years old, who believed that keeping pigeons was a peaceful activity. I met him through a cockfighting friend, Allah Wasaya, or Waso, as he is nicknamed. Waso was put off by pigeon flyers who he considered eccentrics, obsessed with their flocks so much that they could stay on their rooftops, oblivious to the sweltering summer heat. However, despite Waso's disdain of pigeon flyers, both Rafik and Waso were good friends. Rafik ran a stall of *shami* kebab in front of his house from evening until midnight, and Waso was a gourmand who admired Rafik's spicy round kebabs, fried in clarified butter.

Rafik had lived his entire life in a tiny house built by Hindus in Chandni Chowk, in the southern corner of town. Before the Partition, mostly Hindus inhabited the area but after the events of 1947, they migrated en masse, abandoning their businesses and properties for Muslim migrants coming from the Indian state of Haryana. Today, the area is a famous hub of pigeon flyers where one can frequently see flocks of colourful pigeons wheeling and swooping in the air with the soft sound of beating wings. Pigeon flyers are generally not visible from the road, and yet as they perch on the rooftops of their two-storey houses, their whistles are audible only to the trained ear amidst the noise of rickshaws, hawkers, and loud Bollywood music coming from dingy tea shops.

The liminal architecture of pigeon-filled rooftops, as anthropologists Jürgen Wasim Frembgen and Paul Rollier (2014, 78) explain, serves as "an extension of the house, an interstitial zone of familiarity somewhere between the seclusion of the household (*khalwat*) and the exposure (*jalwe men*) of the street." There in the liminal space, pigeon flying enthusiasts like Rafik escape the mundane aspects of daily life and develop an intimacy with their birds to achieve a degree of personal freedom. Such freedom, according to enthusiasts, is enriching as it allows them to achieve a unique and deeply fulfilling experience, and come back energised to the realm of everyday life (Turner 1967, 93–111; 1969, 94–96). Although a pigeon enthusiast perched on his rooftop may seldom look down into the street, he is often engaged in aerial conversations with other pigeon flyers across the rooftops through his birds, a scenario painted most beautifully by novelist Ahmed Ali in *Twilight in Delhi* (1966).

When I first visited Rafik's rooftop accompanied by Waso, we climbed more than twenty steep steps, then went over a wooden rung ladder that led us to the terrace through a hatch. Walls enclosed the rooftop that housed three large pigeon coops made of wood and wire mesh. Rafik was busy flying his pigeons, whistling to them, and waving a blue flag tied to a bamboo stick. His eyes were focused on the birds flying in groups, their shiny wings stretched out under the sun. The pigeons seemed to understand the varied pitches of his whistles and changed their movements accordingly. The first whistle, something like a sharp loud shriek, would send the flock to a distance in a straight line; the second whistle, produced by joining the thumb and index finger, made them change direction; and the third whistle, loud but firm, signalled their return and made them swoop over the rooftop. After some time, Rafik picked up a rusty tin can of grains and, tossing a handful towards the coops, beckoned the birds to come eat by emitting soft sounds of *āoo, āoo* (come, come). All of a sudden, pigeons alighted on the rooftop in a loud flutter of wings. Tending to them, Rafik scattered some more grain and poured water in an earthen pot for the birds to drink. Some pigeons perched on his arm and pecked grains from his hand, while others settled on the enclosing walls, chittering and cooing with pleasure. As the pigeons finished eating and drinking, Rafik guided his birds into their respective coops and closed the doors.

Squatting beside me and Waso, Rafik remarked on the success of his kebabs business "No one can make *shami* kebab like I do in the entire district." Waso nodded in agreement and complimented him by adding, "You must taste his *shami* and see for yourself; he is a true craftsman (*kārīgar*)." Twenty years back Rafik set up the *shami* kebab stall in front of his house and worked every evening, sometimes saving as much as $6 a night. He spent some part of this income on feeding his pigeons and meeting domestic expenditures, the rest would be put away for the impending wedding of his four sons and three daughters. However, life was not always easy. In 1988, when the River Sutlej flooded many villages nearby, young Rafik was working as a tailor, sewing men's *kurtā shalwār*. The increasing competition from other tailors meant that Rafik barely made enough to feed himself and his pigeons. "At the time, pearl millet (*bājrah*) was three Rupees per kilo and I couldn't save that much to feed my pigeons" Rafik recounted. To supplement his income, he set up a shop selling lemon soda in the evenings. The business was good in summer and he was able to save some money to marry his maternal cousin, his childhood sweetheart. However, in winter no one needed a soda bottle and all his savings ran out. Heeding to the advice of some friends, Rafik sold all his pigeons and moved to Lahore and then to Karachi to work in garment factories. However, there in *pardees* (foreign land), he felt sad (*dil naī lagtā thā*) without his wife and pigeons. So Rafik moved back, repurchased all his pigeons, and started working as a vegetable vendor. After six months of hardship, he gave up. "That was the time when I longed for every single Rupee," Rafik reminisced, and then "one day, sitting near my hungry pigeons, an idea struck me and I set up the *shami* kebab stall, going to bed with a profit of thirty-five Rupees the first night, and with fifty Rupees the second night." Since then, Rafik has only seen success.

"I keep pigeons for my *shauq* and Allah supports me in sustaining it," Rafik argued, recalling times when he sat on the floor with a sewing machine, bent double almost all day, that gave him chronic pains in his body. But now, his *shauq* of flying pigeons has relieved him of all his troubles, "this is my only recreational activity (*tafreeh*) and a type of exercise that keeps me fit." Looking at his pigeons who were now peacefully settled in their designated places within their coops, Rafik emphasised how pigeons are his most valued property (*malkeat*), and how he raises them "like children." He suggested that the birds have helped him find meaning to life, and connected him to other pigeon flyers who share his *shauq* and appreciate his deep commitment to pigeons. His birds also helped him connect to the ways of life of his pre-Partition ancestors, and their presence in the house motivated him to find success in his work.

How does this more-than-human sociality provide a sense of fulfilment and freedom to pigeon flyers and enable them to counteract life problems? And how does the *shauq* of keeping and flying pigeons animate a flyer's social life, including his prioritisation of time and care, organisation of work, development of social relationships, and accomplishment of honour and status? Rafik's life story, and the stories of other *shauqeen* pigeon flyers in rural Pakistan reveal that pigeon flying is not an ordinary pastime, rather it structures the lives of pigeon flyers and shapes their understanding of the self and others. Unlike the past when this practice was the preserve of the princes and kings in South Asia (see Chapter 1), in contemporary South Punjab, many pigeon flyers I met were poor men, working as farm labours, milkmen, shopkeepers, coachmen, and rickshaw drivers, although some were from the educated middle-class and worked as school teachers and clerks. Yet, despite often difficult financial circumstances, many viewed their *shauq* as one that affords them a respite from the mundane, and allows them to compete and win *izzat* among their peers. By discussing the life stories of pigeon flyers, I argue that the men's relationship with their pigeons is centred around the protracted care of the birds and allows them to construct a social universe where they develop new relationships (as master and disciple, or friendships with other flyers), and achieve symbolic gains (*izzat*) by fulfilling their personal enthusiasm (*shauq*).

Symbolically, the pigeon has long been associated with peace, especially in Judeo-Christian and Islamic religions. This tradition can be traced to the famous tale of Noah's dove whose flight signified a new era of hope, peace, and prosperity. Yet, as Jennifer Price (1999) reminds us through her discussion of the extinction of the passenger pigeon, this bird of peace has been killed on a massive scale in more recent times. But equally, this ubiquitous bird is found in city squares, on rooftop coops, and sometimes in million-dollar pigeon races, and continues to be a part of our social experience. Sociologist Colin Jerolmack's impressive ethnography *The Global Pigeon* (2013), traces the pigeons' prints on human social life in New York, Berlin, London, Venice, and in Sun City. The "cross-species encounters" developed through pigeons, Jerolmack argues, weave into "people's social worlds" and allow them to achieve unique experiences (2013, 13–14). In this way, pigeons become what Molly Mullin (1999) suggests are "mirrors and windows," a symbolic resource to connect humans and structure their social relationships and lifeworlds.

In rural Pakistan, pigeons are more than just symbols for enthusiastic flyers. The birds, as Rafik affirms, possess individualistic traits and a distinct personality. This means that to develop a deep and embodied understanding of pigeons, a flyer must develop *shauq* to understand the self in connection with pigeons and take his avian companions as, in the words of John Knight (2005, 1) "subjects rather than objects, … as parts of human society." Therefore, pigeons are not simply symbols through which enthusiasts structure their lifeworlds; they are social actors, cherished companions, intimate beings, and active agents who share the days and nights of their flyer and influence his status and standing among his peers.

In the rest of this chapter, I will first look at the *ḍarī* pigeon flying competition and discuss how its structure allows *shauq* and work to coexist. I show why a relationship of intimate labour and care of the bird is crucial for developing and maintaining this *shauq*. Then I will look at the *khokhā* pigeon flying competition and examine how pigeon enthusiasts develop friendships and hostilities and engage in pigeon wars. Towards the end, I will explore the master and disciple relationship, particularly by examining the *asmānī* pigeon flying competition where pigeon masters are treated with utmost respect. "Because sport is fun," anthropologist Joseph Alter (1992, 19) argues while writing about

Indian wrestling, "does not in any way mean that it is insignificant. As a cultural system, sport is meaningful to the same extent as systems of production and exchange." Although pigeon flying in South Punjab is rarely described as a sport (*khel or kheḍ*), yet I suggest that its close examination showcases the values attached to it (such as care, respect, time-management, protection, socialisation, and hard work) and explicates the concept of honour and prestige linked to the wider formulation of masculinity in everyday social relations among the enthusiasts.[2]

Shauq and work: escaping worldly troubles through *Ḍarī* competition

Pigeon flyers employ the recurring term *shauq* to demonstrate their longing, passion, and dedication to keep and fly pigeons. More specifically, the word *shauq* embeds the notions of obligation and responsibility to care for your flock and to protect it from predators such as cats and hawks. It also gives the pigeons the position of a subject with particular likes and dislikes, and leads a flyer to understand the pigeons needs to develop mastery over them. This *shauq* is aptly captured in the local term "*kabūtar bāzī*." The word *kabūtar* means "a pigeon," while *bāzī* generally connotes "excessive habit": one that entraps people and blinds them to other aspects of social life. The suffix *bāzī* in South Asia, as Akhil Katyal (2013, 475) explains, is "a common expressive idiom of heightening the usual, of playing a game, of cherishing a sociable overindulgence with certain objects (alcohol, *darubaazi*), practices (dancing, *naachbazi*) or people (prostitutes, *randibazi*)." In the *shauq* of pigeon flying, the word *bāzī* means to fly the bird "with devotion and amusement," and sometimes "to bet on pigeons" (*kabūtar te bāzī lāwna*). However, among the general public, the term *kabūtar bāzī* is often used in a pejorative sense, a passion with the potential of growing into a dangerous obsession. Local people and close family members of animal keepers, especially their wives, argue that *kabūtar bāzī* lures men towards an imaginary utopia and diverts their attention away from real work. The effects of *shauq* on family members is the subject of another chapter (see Chapter 5), yet here it leads us to question the inter-relationship between *shauq* of *kabūtar bāzī* and work, particularly, how *shauq* becomes a type of work; why work is performed on a daily basis to maintain *shauq*, and how *shauq* leads to achieve personal well-being and fulfilment despite poverty.

Like many other flyers of the area, my friend and pigeon flying master, Ustad Qamar-ud-Din starts his day by opening the doors of the pigeon coops on his rooftop. A middle-aged man of short height, he has lived all his life in the village of Shahpur, twelve kilometres south of Kahror Pacca, where people know him as Qamo. The silver amulet tied in black thread around his neck protects him from the evil eye of jealous people and his black Sindhi cap and recently grown beard indicate his piety. Flying his flock of 100–150 pigeons about twenty to thirty metres above ground and over the neighbouring houses, Qamo engages in *ḍarī* pigeon flying competition every

Figure 2.1 Qamo with his pigeons, October 2008.

Figure 2.2 Qamo with his pigeons, July 2017.

morning and afternoon. Like other flyers of the area, he uses a *dogazā* (a vertical net) and a *jhaṇḍī* (a flag tied to four-foot bamboo pole) to direct the flight of his flock, guiding them through whistles to fly in a tight bundle or merge with the opponent flock. As the flocks met in the air, Qamo showed me how to cease whistling and summon the birds back by shaking the tin can of grains, tossing up *kuṭṭī* (a pigeon with a clipped wing who flutters back to the coop),[3] and throwing some grains into the air.

"Do you know when one becomes a *kabūtar bāz*?" Qamo asked me one morning. He answered his own question by adding, "when he starts recognising the opponent's birds in his flock." This is not easy at all. To identify a single pigeon in a flock of 100–150 birds circling high up in the sky and constantly moving requires true mastery, including detailed knowledge of each pigeon's bodily features and flight pattern. But almost all flyers can do this easily. They can immediately tell which pigeon joined or left a flock, why that pigeon did so, and how to lure that pigeon back. Over the period of a month, Qamo sometimes captures up to twelve pigeons of his opponents. He then evaluates the captured pigeons according to their colour and pattern and either exchanges them with another flyer or, if they are of good breed, mates them with pigeons from his flock and trains them as his own.

Pigeons in a *ḍarī* competition take off in the early morning and in late afternoon to avoid the intense heat and glaring mid-day sun, and also so that flyers can attend to their livelihood. When I first met Qamo in 2008, he had a horse-cart to carry people from the village to the city and, till 2005, he made as much as $5 on a lucky day. Starting the day with a two-hour session of flying pigeons, Qamo would then offer feed and water to his horse and carefully clean the shiny blue leather seat of the cart. Playing the cassette of a famous Seraiki singer, Attaullah Esakhelvi, on his old Panasonic cassette player, Qamo would wait for customers near the village hospital along the road that led straight to the town of Kahror Pacca. However, after 2005, a surge in motor rickshaws brought a sudden decline to Qamo's business. In 2008, Qamo had to wait for hours to get even six passengers and sometimes "only made one round trip the entire day." To add to this, the price of fodder kept going up and the horse was getting older. Amidst all his financial troubles, Qamo still found time to bring his day to an end by flying pigeons for two hours before closing the door of the coops at sunset.

When we met again in 2014, Qamo told me about the tragic death of his horse. He now kept a donkey and used the donkey-cart for transportation of goods to and from town that fetched him less than $2.50 a day. His routine, however, did not change. He flew his pigeons in the morning and then transported ice blocks, burnt bricks, or sometimes sacks of fertilisers or fodder, ending the day with another round of pigeon flying on his rooftop. Although the horse was replaced by the donkey, his interest in keeping pigeons remained the same over the years. Whereas the horse or donkey were a means of livelihood, pigeons were kept for *shauq*. From his meagre earnings, around one-third was spent on feeding pigeons expensive seeds. Qamo's decision to invest a major part of his earning on the pigeons was crucial because, as he told me, the birds were the primary source of his pleasure. An interesting point to note is the distinction Qamo made between his work and his *shauq*, both of which involved animals. Pigeon flying brought him joy and required constant attentiveness and care; the horse and the donkey were a means to achieve that end.

In his book *The Conquest of Happiness* (1932, 211), Bertrand Russell argues that "interesting work" requires "the exercise of skill" to a degree that a person keeps improving it and develops the mastery to outwit a skilled opponent. This then becomes a source of pleasure, delivers the satisfaction of a higher order, and provides "continuity of purpose" (1932, 210). However, when a person can no longer improve in skill, the work becomes repetitive, unpleasant, and uninteresting. Qamo's relationship with his horse or donkey was important, since the animals were his source of livelihood but the repetitive work lacked improvisation, innovation, and improvement in skill. Whereas the universe of pigeon flying was vast and required a deep knowledge of breeding, feeding, curing, and training pigeons; identifying pigeons through their colours and patterns; flying the birds in various formations at different heights and over various distances; and interacting with other flyers on almost a daily basis. While whistling with two fingers in his mouth or weaving the flag anxiously, Qamo's day on the rooftop was an adventure, a way to learn new strategies of controlling the flock in the air and artfully making them lure opponent pigeons. This absorbing *shauq* of keeping pigeons was important, interesting, and valuable.

Yet *shauq*, Qamo told me, is not only to fly and catch pigeons but to develop a "two-sided relationship" (*do-tarfah rishtah*) with the bird. The relationship is reciprocal because pigeons amuse their keeper, while in exchange they receive food and protection from predators such as cats, mongoose, snakes, and goshawks. However, the development of this relationship involves the mutual acceptance of each other, as Qamo asked: "Have you ever seen a person flying a flock of crows or parrots?" The general idea is that, unlike other birds, the pigeon is amenable to domestication. The key is reciprocity, as Qamo remarked, "I trust my pigeons and they trust me." He said he believes his pigeons would never go astray while flying in the air while the pigeons understand that he would never stop caring for them "like a mother."

The maternal love emerging through this relationship built on reciprocity and mutual acceptance enabled the men to cultivate what Radhika Govindrajan (2018, 44–47) describes as "gendered relatedness." Qamo and many other flyers demonstrated this gendered relatedness by engaging in activities that are not usually considered in the male domain in rural South Punjab, such as cleaning pigeon coops, carefully selecting the diet for the birds, decorating pigeons with anklets and colourful beads, and protecting them from harsh weather. It was also this gendered relatedness that enabled many keepers to develop a detailed knowledge of each pigeon's biographical information: their age, the history of their illnesses, their flying techniques, weaknesses and strengths, and in which season their flying performance was likely to decline. This intimate knowledge of pigeons, according to Khuda Bux, was also critical in selecting a pigeon's mating partner.

In the early months of my fieldwork in 2008, I came to know Khuda Bux, a chiselled-face pigeon flyer in his mid-40s. Khudla, as he was locally known, was an amateur singer and pigeon enthusiast who had lived his entire life in a one-room adobe house as the oldest bachelor of the village. He worked as the livestock keeper for a rich farmer and, except for flying pigeons in the morning and in the afternoon, spent all day cutting and manually chopping the fodder, feeding and milking the cattle, bathing them after cleaning their shed, and removing the manure. When we met again in 2014, Khudla was married and worked at the farm of the village landlord on a monthly salary of 7000 Rupees (about $44) and two sacks of wheat during the harvest season as his yearly bonus. The employment was good and Khudla found a fine balance between his work and *shauq* until, in 2017, the landlord decided to move to Bahawalpur to educate his children in an English-medium private school. The farmer who acquired the landlord's arable land on a yearly renewable lease decided to engage his extended family for menial labour, which left Khudla unemployed. "I looked for work everywhere, went as far as Hitar (a place across the River Sutlej) to work as a labourer. A man has to feed his family, and I had to feed my pigeons too" Khudla recalled that "harsh" time. He looked for work for many more months until a friend offered him two cows on contract that meant Khulda could sell the milk of the cows and live off of that but every second new-born calf was promised to the original owner. The precarious nature of work, months of unemployment, and the anxiety of finding a livelihood did not shake Khudla's *shauq* to fly pigeons and, like Rafik and Qamo, he described it as respite from his troubles. "Whenever I worried too much about money, I would go to the rooftop and sit close to my pigeons," Khulda told me. The presence of pigeons, he said, raised his sense of responsibility and, as the birds needed grain, he did not spend a single day sitting idle in the house.

From close observation, Khudla came upon a "secret" mating pattern of the birds that many flyers did not know about. "Pigeons are monogamous creatures, so the mating pair has to be selected carefully," he responded when I asked him about breeding high-quality pigeons. He believed in the crucial importance of a female pigeon who, he said, affects the quality and characteristics of the chick. In addition, the generational relationship between male and female pigeons should be taken into account before pairing. Lastly, he suggested, the physical appearance (colours and patterns) of the mating pair need to be taken into consideration to craft a unique chick. Khudla knew that a fine-bred pigeon is highly valued among other flyers and that it reflects well on the keeper's *izzat*, so he came up with the following breeding pattern and choices.

Interestingly, all Khudla's preferred mating choices for pigeons involve an incestuous relationship that is taboo among people in South Punjab. In figure A, the female pigeon is mated with her paternal uncle (*chachā*) which is opposite in figure B, where a male pigeon is selected to breed with his paternal aunt (*phuphī*). In figure C, there is cross-generational mating between siblings.

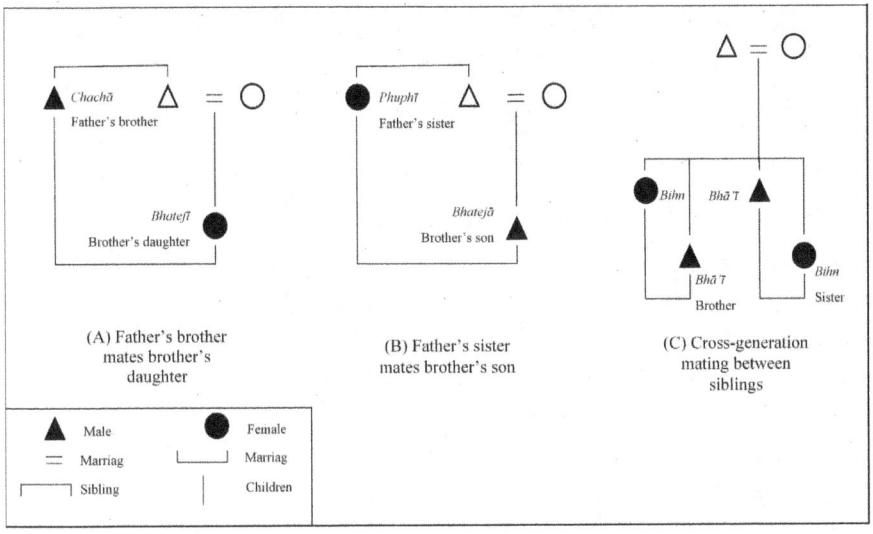

Figure 2.3 Preferred pigeon-breeding choices for Khudla.

We can analyse these mating preferences for pigeons as a part of Khudla's unconscious structure, established in binary opposition to accepted human mating choices (Levi-Strauss 1963). Yet, such an analysis would overlook human–pigeon sociality developed through sharing a common space over an extended period of time. While living in close proximity, humans and pigeons develop an extended knowledge of each other—of body, smell, touch, sound, and behaviour—and form a connection that is personal, valued, and embodied. This was such a personal and embodied connection based on extended knowledge of the self and the other that enabled Khudla to develop unique breeding and mating patterns.

Generally, Khudla and other flyers preferred two breeding methods: breeding in a separate cote "private *kamrah*" (or private room), and breeding in a dovecote (*khuḍā*). The first method is designed for newly caught stray pigeons for them to become accustomed to the coop and to their new keeper. It was meant to give them a "honeymoon" environment with plenty of grain, a partner, and privacy from other pigeons. The second method of breeding was meant to increase the numbers in your flock. From the dovecote, the best chicks are recruited into the flock while stragglers are often sold in the market.

The first time I followed Rafik into his dovecote, I experienced a stuffy, dark, and dingy environment. Rafik rented the room from his friend at 1000 Rupees (about $6) per month, specifically to breed pigeons. This small room at the corner of the main bazaar, full of a pungent stench and the foul odour of pigeons' dung and hatching, housed almost thirty pairs that were spread

Figure 2.4 A pair of newly hatched chicks.

across the ground in small boxes made of wood, in brick-made shelves, and near the door. "This is my favourite place, where I spend up to seven or eight hours daily," Rafik told me, as I strained to hear him above the noise of rickshaws, honking motorbikes, and high-pitched hawkers. There, he said, he scrutinised the breeding pattern of pigeons and checked on the condition of the eggs before neatly making notes on each pigeon's birth date and time.

I wondered at Rafik, Qamo, and Khudla who, instead of staying home with their family, spent their precious free hours breeding pigeons. The close association that they had developed with their birds transcended species boundaries, and made pigeons a part of their kinship circle. Many flyers had developed a feminised form of love for their birds, or what they referred to as *mān alā payār* (maternal love). They spoke about the intensity of their love in crafted proverbs like wearing *shauq deān wangān* or the bangles of *shauq*. Elsewhere, Kirin Narayan and I discussed the use of such proverbs by men in the South Asian context that carried different meanings, including how *shauq* made them feel feminised, shackled them to responsibilities, and added beauty and pleasure to everyday life (Narayan and Kavesh 2019, 8). Here, I emphasise that the use of such proverbs defended the men's choice of spending an inordinate amount of time with their birds, and justified their involvement in gendered care.

Sitting around a fire on a chilly night in December, my pigeon flyer friend Aslam shared how *shauq*, the passionate predilection that adorned his life, sometimes made him act outside of culturally defined gendered roles.

Once a female pigeon (*kabūtarī*) died, and I raised her chick like a mother. Every day, I fed him on milk with a dropper and with chewed grains. At night, he slept near me, on my charpoy in a carton box, so I could ensure he stayed warm in the cold weather. He soon became my favourite pigeon.

After three years, when he died due to a digestive problem (*un-pach*), I could not eat for two days from grief.

Aslam's story illustrates the intense gendered care and bond created between a flyer and his birds, much like a bond between a child and a mother in rural South Punjab. However, such "maternal love" towards one's pigeons is sometimes interpreted as an effeminate obsession by village people. In South Punjab, men are not expected to display what are commonly seen as non-masculine traits such as care giving and tenderness, the expression of affection or grief, or marvelling at aesthetics. As we will see in the discussion on cockfighting and dogfighting in coming chapters, men are generally expected to maintain bodily and emotional composure. From childhood, transgressions are monitored and an embodied habitus is carefully cultivated. However, in the context of pigeon flying, it appears that the men disregard such socialisation to a great extent. In the privacy of their rooftops, they commonly express affection towards their pigeons and appreciate their colours, patterns, and movements. Some men, like Aslam, even shed tears over the loss of a prized bird. Perhaps it is the architecture of the roof—seclusion from the social world that offers a liminal space for non-normative expression of care—or total immersion in the activity that enables such transgressions. It seems that the term *shauq* is especially suggestive in this context, as one of its characteristics is a type of uncontrollable obsession, effectively rendering non-masculine behaviour acceptable.

Speaking about such gendered affection, Ustad Nadir Abbass once remarked, "I care for my pigeons as a woman cares for her children." A retired school teacher, Nadir Abbass was in his late 50s when I met him, and famous as a respected master (*ustād*) among pigeon flyers. This attachment, he argued, made him appear feminised to others, yet Connell reminds us that masculinity is not a fixed personality trait of a man, rather, there are multiple masculinities that are foregrounded according to contexts (1995, 76–77). For instance, as the discussion in Chapter 5 suggests, South Punjabi men possess a range of masculinities, and while pigeon flyers felt at ease in expressing what might be considered largely feminine traits in the context of rural South Punjab, outside of that milieu, they adhered to the hegemonic gender model that valued tough, competitive, inflexible men. It is vital to note that these are essentialised gender distinctions, and are subscribed to different degrees as culturally meaningful approximations. Still, the gendered attitude the flyers displayed in the company of pigeons was unique and rarely seen in other settings. Taking a pigeon in his hands or feeling the softness of her body, stretching her wings or cheerfully releasing her in the air, Ustad Nadir Abbass used to remark: "In the love of pigeons, I forget everything else."

The intimate labour and care allowed most pigeon flyers to develop a deep sociality with their birds and acquire a sense of happiness, fulfilment, and satisfaction despite living in grinding poverty. The men took inspiration from their *shauq* to look for work and fight their worsening economic condition.

In her discussion on the Ansari weavers of Banaras, Nita Kumar (1989, 156) highlights the subtle relationship between *shauq* and work, and suggests that *shauq* provides a degree of freedom from the demanding work routine. Similarly, Clare Wilkinson notes that *shauq* becomes crucial in fighting the troubles of life (or *pareshani*) and motivates women embroiderers of Lucknow to acquire superior skills, "Only pure *shauq* can stimulate learning and only *shauq* can explain facility in the craft" (1999, 133). The *shauq* of flying pigeons enables people to achieve, as Whitney Azoy suggests while discussing the *shauq* of Buzkashi in Afghanistan,[4] a "sense of perceived separateness from the serious stuff of day-to-day reality" (1982, 8). This reality is not just physical, it is also social and involves class hierarchy, sectarian affiliation, ethnic belonging, and linguistic differentiation. However, in pigeon flying, *shauq* transcends these ascribed affiliations, leaving everyone to be judged and assessed on their association with pigeons and the seriousness of their passion (Narayan and Kavesh 2019). This is where *shauq* of flying pigeons intersects with *izzat* and transforms this activity into a fight for honour among peers.

Pigeon wars: winning honour through *Khokhā* competitions

I first met Aslam in August 2008, the month when Pakistan's military ruler General Musharraf resigned as president of Pakistan, putting an end to almost nine years in power, and I remember how we both digressed from the topic of pigeons to Musharraf's resignation and spent hours discussing the future of Pakistani politics. A skinny, dishevelled man, Aslam was an expert craftsman of delicious local sweets including *motichoor ladoo, jalebī, toshā, barfī*, and sold them in a small shed at the outskirts of Kahror Pacca. He lived with his wife and four sons as a part of the extended family in an adobe house just on the edge of the main canal, whose milk tea-colour water separated the area from north-west villages. When I met him again in 2014, his life had changed. After an unexpected quarrel with his younger brother, Aslam left the family-owned sweet shop and tried other kinds of work, eventually settling as a woodcutter. His two sons, who had assisted him in preparing the sweets, were now going to a local school. What remained consistent throughout those years was Aslam's *shauq* of keeping and flying pigeons. Despite sudden changes in employment, Aslam maintained a deep attachment to this activity that gave him personal gratification and helped him maintain a close relationship with other flyers.

The relationship between two pigeon flyers is either one of friendship (*yārī*) or that of hostility (*mukhālfat*). In a friendly relationship or in a state of peace (*sulah*), the men exchange pigeons, train birds together, and provide each other with help and advice. The men use the word *yār* (friend) instead of *dost* (also meaning friend) to signal a close tie based on the principles of affective support, cooperation, and teamwork (Narayan and Kavesh 2019). Such friends may offer moral and even financial support in times of need, and emphasise the importance of remaining sincere and trustworthy through difficult times. A pigeon flyer's success reflects his friend's success; when he wins, his trusted friends rejoice with him, just as they sympathise when he suffers a loss. Informal gatherings play an important role in fostering this friendly relationship. Draping their arms around each other's shoulders, sitting together in a tea stall, visiting each other's coop, or playing cards or dice, many men use famous Seraiki proverbs such as "*yārī ty sab qurbān*" (everything may be sacrificed in friendship) or "*yārāṇ nāl bahārāṇ*" (only with friends is there springtime) to signal the depth of the relationship (Narayan and Kavesh 2019).

Yet, all pigeon flyers are not friends and some are opponents in a declared state of war (*jang*). While reading Marius Kociejowski's *The Pigeon Wars of Damascus* (2010, 37), I noticed remarkable similarities between the symbolism of war used by Syrian pigeon fanciers and by rural Pakistani flyers. Such symbolism is noticeable again in the Aljazeera documentary *Pigeon Battles of Cairo: Egypt's High-Flying Sport* where the protagonist, Koka, uses his pigeons to win honour in his last "fight" against one of Cairo's esteemed pigeon flyer. In both cases, the symbolism of war is expressive and forces pigeon enthusiasts to accumulate honour through the cherished companions of their social universe: pigeons.

When Aslam left his sweet shop after the quarrel with his brother, his friend Balo informed him of a new pigeon competition, a *khokhā* competition, where the symbolism of war is most explicit. *Khokhā* (literally a coop)

Figure 2.5 *Yār* pigeon flyers (photo by Mahar Akmal).

competition also involves enticing opponent pigeons by tossing up grain, however, it does not take place on rooftops but on the ground. Before the competition, both pigeon flyers agree upon a date, time, and venue, as well as the number of participating pigeons, and set a sum of money to return captured pigeons. On the pre-decided afternoon, just after Asr prayer, the competition is staged in open grounds in front of a small audience. The selection of a neutral place is considered crucial not only because of home-support but, as Aslam noted, "Because the pigeons can remember the place, direction, and surroundings, we chose a neutral place so no flock gets an advantage over the other." The flyers, accompanied by their close friends and supporters, lug their coops on a donkey-cart or a motorbike-cart and place it in front of each other at a distance of about thirty metres. The ground in front of each *khokhā* is cleaned and showered with water, a practice aimed at keeping the pigeons happy so that they fly better. As flocks of about 100 pigeons from each side take flight and coalesce in the air, flyers summon them back to their coops, instantly identifying any opponent pigeon who has mistakenly landed with their flock, and try to seize the pigeon in a net or by hand. This process continues for almost two hours until sunset, with the flyers trying to capture the opponents' birds. Captured pigeons are kept separately until the end of the competition when both pigeon flyers trade captured birds for the amount of money initially agreed upon.

I was never invited to participate in a *khokhā* competition, the reason was, as Aslam, Balo, and other flyers told me, that I lacked the knowledge of differentiating pigeons through their colours and patterns. This is crucial because it enables enthusiasts to instantly and instinctively identify a foreign pigeon in their flock and capture it. Standing there among the audience, I realised that I was not a true *sahuqeen*. A true *sahuqeen* was expected to easily identify,

Figure 2.6 Khokhā competition, 2008.

following John Galaty's (1989, 225–226) categorisation developed in the company of Maasai pastoralists: (a) his own pigeon in the opponent's flock, (b) the opponent's pigeon in his flock, and (c) any missing pigeon within his flock.[5] Since all the men present, including the young boys assisting them, possessed this remarkable ability to identify pigeons amidst all the whistles, dust, and flapping wings, I started exploring the pattern to this identification process, largely by noting pigeons' colours and other external features.

A pigeon's colour is the major indicator for categorising and evaluating the bird. There are three broad categories: *bad-rang* (mixtures of different colours), *bhore kābre* (dappled), and *shirāzī* or *rang-dār* (colourful).[6] The first two categories are found in every flock and are considered common, whereas *shirāzī* pigeons are, as Rafik assured me, "all about beauty." The principal colours of *shirāzī* pigeons are six: *lāl* (rusty red), *sunehrī* (rose gold), *kāsnā* (light grey), *pelā* or *zard* (arylide yellow), *sabz* or *sabat* (light green), and *syāh* (black). Among all colours, *lāl* is the most treasured and men express their pride in possessing *lāl* pigeons. About the *lāl* pigeon, they say "*rang barā qemtī he*" (this colour is very precious), or "*akhen ko bhāndā he*" (the eyes fancy it). Many flyers appreciate the shine in the colour of *lāl* pigeons and the joy of watching them fly high in the skies among the flock with distinct elegance.[7]

Lāl and other *shirāzī* pigeons also come in various patterns. The most favoured pattern is *cap*, which indicates a single white wing. If a pigeon is *lāl* and has a white wing, it will be referred to as *lāl-cap* (see below figure). The second important pattern is *pāmoz* (feathers on a pigeon's ankle), then *cotī* (a crest), *togah* (a white tail), followed by many others. Each new pattern adds to the name of pigeons, for instance, *lāl-cap*, *lāl-cap-togah*, *lāl-cap-togah-pāmoz*, and so forth. The point to note here is that the actual numbers of lexemes of *shirāzī* colours are limited; however, when combined with other lexemes of the pigeons' pattern, hundreds of names are produced. These names help enthusiasts identify a particular bird in the flock of more than 100 pigeons, provide individualised care to their birds, and develop a deep more-than-human sociality with them.

Pigeons with unique colours and patterns have good market value and every additional pattern influences their price. For instance, Aslam remarked that *do-cappah* (two wings of different colours, other than white, such as one red and one grey wing, or one light green and one black wing) is considered the most valuable *shirāzī* pigeon and the price of such rare pigeons can easily go up to 400,000 Rupees ($2500). Even though many flyers experiment with breeding patterns to produce a *do-cappah*, I never saw one during my fieldwork. In addition to distinctive colours and patterns, pigeons' salient features (*sifat*) and mismarks (*'aeb*) affect their market value. For instance, if a *lāl* pigeon has a particular shine to her colour and possesses a white beak and a long neck, the bird will be highly valued. Conversely, the value of a pigeon may decrease due to mismarks such as having a few colourful feathers or a speckled chest. Although a good colour *shirāzī* pigeon can be bought for about \$6–\$9, Saifal remarked that finding a pigeon with good features and fewer mismarks is truly a challenge.

Saif-ur-Rehman was a small farmer who grew cotton, wheat, and barley on his three-acre farm and what he made from it went mostly into his *shauq*. Nicknamed Saifal, he was a short man with middle-parted hair who had lived his entire life in Village Seekrain with his wife, two sons, one daughter, mother, and a widowed paternal uncle. Like Aslam, Saifal developed a keen interest in *khokhā* competitions. He believed that, although all pigeons have imperfections, over the years he was able to breed and keep those with few mismarks. Such skills had brought him honour among peers who regarded him as an expert in selecting quality pigeons. He told me that when he opens the doors of his *khokhā* in front of opponents, the quality of his pigeons intimidates them.

Colours and physical features are markers of a pigeon's natural beauty; however, the men fix anklets (*ghungro*) to the pigeons' feet to further enhance their appearance. While staying with Qamo on his rooftop, I learnt how mesmerising the activity of keeping pigeons can be. He would fly his pigeons at daybreak and the sound of fluttering wings and the cooing of pigeons would blend with the metallic music of their anklets to create a feeling of peace. However, not all pigeon flyers use anklets, some fix distinctly coloured beads (*motī*) instead. Saifal, for example, told me that beads work as his trademark and distinguish his pigeons from others. For instance, his pigeons had tricolour beads—three rings of orange, black, and green beads—and other

Figure 2.7 Rehan Baig showing his collection of pigeons. "A" is a kāsnā-kawā (a red-eye, grey-colour pigeon), while "B" is a sufaīd kawā tie-ālā (white with a tie – symbolically indicating the shape of feathers on the chest). The pigeon in "Figure C" is the most cherished one: a precious lāl-cap (one rusty red and one white wing).

flyers in his area were aware of this. When he participated in a *khokhā* competition, the beads, along with the distinct colour of his pigeons, would assist his supporters in identifying his birds.

The selection of uniquely coloured pigeons with distinct features and decorating the birds with beads demonstrates a flyer's aesthetic values and his *shauq*. It bolsters his *izzat* among peers and elevates his symbolic capital. However, if the elegantly decorated pigeons cannot fly well during the *khokhā* war competition, they demonstrate a weakness in the flyer's abilities. A pigeon flyer, according to Saifal, must invest time and resources in his *shauq* and epitomises his commitment to, and knowledge of the pigeons' body by training them to fly like champions. As I have argued, there are different expectations required of the flyer, for instance, he must adhere to a demanding daily regimen, be able to develop skills and strategies associated with breeding, possess mastery over a wide lexicon deployed in the craft of rearing pigeons, become competent in the art of whistling, and show an ongoing commitment to the activity. Another crucial skill is the technique of training pigeons to fly within a close cluster, since a well-trained flock is the main source of a pigeon flyer's masculine *izzat*.

The training of pigeons is complemented with reward (*pyār*) and punishment (*mār*). In *pyār* (literally, to love something or somebody), the flyers kiss their birds in affection, feed them diced almonds and pistachio by hand, and gently stroke their heads. Whereas *mār* (literally to beat somebody) means to stop giving grain to a wayward pigeon, or hit the pigeon (with a stick) "as gently as we hit our children to teach them a lesson," Rafik once told me. Such care involving reward and punishment means, as anthropologist Dimitrios Theodossopoulos (2005, 22) notes while discussing the stringent farm rules of the Vassilikiot farmers of Greece, to ascribe human values to non-humans to make them comply with human preferences. For pigeon flyers, such training practice is required if the bird is to be transformed into a *girdān*: a well-trained pigeon. The *girdān* pigeons are leaders of a coherently functioning flock. They can be male or female and are considered loyal birds who understand their flyer's orders and comply with his instructions. They can fly long distances, dip and rise in response to the flyer's whistles, fly in various formations, and always remember the location of their coop. Aslam claimed that *girdān* pigeons can recognise the colour of their flyer's clothes from a distance, the colour and shape of their coop, and even the colour of the tin can grain box. For Saifal, *girdān* pigeons possess the remarkable ability to remember directions and, no matter how much grain the opponent flyer throws to entice them, they always land on their own coop. It is easy to turn juvenile pigeons into *girdān* as they are amenable to training and less conditioned, however, "the real mastery," Chacha Munda emphasised, "lies in catching someone else's *girdān* and turning it into your own *girdān* through extensive training."

Not many people in the village knew that Chacha Munda's birth name was Munawar Shah. He was a jolly man in his 60s, broad-shouldered with a short-beard, who always preferred the company of teenagers. His choice of colouring his beard with henna and wearing embroidered kurtas made him famous as a *mundā*, Punjabi for a boy. Chacha Munda's (literally, uncle boy) technique of turning a pigeon into a *girdān* was based on strict training. To prepare for competing sports, as Rebecca Cassidy notes while discussing thoroughbred horse racing in Newmarket (2002, 114) and as Loïc Wacquant shows while explicating the experiences of Chicago-based boxers (2004, 70), adherence to a strict routine of training is crucial. Chacha Munda also followed this rule and climbed up on his rooftop at noon, a time when most of his rival flyers were not active. In the first stage, he would fly the captured pigeon with his own flock for a few days and provide him with a substantial amount of grain. After a week, he would move on to the next stage and ask a friend to try to lure the pigeon with grains. If the pigeon landed on the friend's rooftop, the friend was supposed to hit him (*mār*) with *dogazā* (the pigeon net) and release him to return to Chacha Munda's coop where the pigeon received more punishment by being denied grain for that night. Chacha Munda believed that this training process repeated a couple of times made the bird aware of why not to land on a foreign rooftop and remain loyal to the flock. Some intelligent pigeons learned quickly and became *girdān* with time.

Punishing the pigeon was not an act of cruelty, but part of the necessary process to strengthen the bond between the flyer and his birds. Chacha Munda thought it was for the greater good of the pigeons (*kabūtaron dī bhalāī*) as they learnt the basic principle of loyalty. Training a pigeon to remain loyal to the flock, he argued, was as crucial as teaching a child to respect elders, eat with the right hand, or avoid talking to strangers. Deviation from basic etiquette could result in a child getting *mār* from parents, extended family members, or even teachers in schools or madrassas. Therefore, when Chacha Munda decided to punish his pigeons, he displayed a sort of parental care. This was reiterated by many flyers when discussing the topic of *mār*, all of whom equated pigeons with children, and suggested that punishment is central to the human–pigeon relationship as it enabled a trained pigeon to develop an understanding of the preferences of the flyer and respond to his instructions.

In contrast to punishment, the *pyār* or love for pigeons becomes explicit as the keeper prepared the diet of the pigeons with care. On average, a flock of 150 pigeons eats two kilograms of millet per day (equal to $1), which is first washed to remove dust and then dried under sunshade to preserve its vitality. Different other products supplement millet over the seasons. For instance, "hot" diets such as egg, *ghī* (clarified butter), almonds, pistachio, and honey are used for the winter season to keep pigeons healthy and active. In summer, a combination of "cold" diets such as butter, wheat grains, and fennel seeds are used to maintain a freshness among the flock. The categorisation of food into "hot" and "cold," as anthropologists have noted, is deeply ingrained in South Asian culinary patterns and influences people's everyday dietary choices.[8] While writing about North Indian wrestlers, Alter (1992) noted that *ghī* and almonds are used in different seasons by the wrestlers to increase the general strength and stamina. Among rural Pakistani pigeon flyers, I also recorded similar observations about *ghī* and almonds. Such overlap in the dietary choices of humans and pigeons blurs the line that distinguishes human and animal, culture and nature, and self and other in rural South Punjab. Also, such choices reflect the affection and *pyār* of the flyer towards his birds, and sustain his *shauq*.

Another expression of the flyer's *pyār*, care, and responsibility towards his pigeons is expressed in curing the bird. Very few flyers consult veterinary doctors for the treatment of ailing birds; instead, most of them take their pigeons to the *ustād* (masters) or treat them personally by using a combination of traditional and biomedical therapeutic techniques. The ability to heal a bird displays the skill and competence of the flyer whose intimate knowledge of pigeons' physiology and common pathologies earns him *izzat* among his peer. Such knowledge is carefully guarded and passed from the *ustāds* to his most dedicated disciples. In dire circumstances, when a pigeon cannot recover from illness or injury, he is killed to relieve him of further misery. Killing (*hath rakhna*, literally putting a hand) in this situation is considered another responsibility, the last duty (*farẓ*) from the man to his cherished animal.

As the flyer cures, feeds, and trains his pigeons, and provides them attention, time, and care, he carefully nurtures a more-than-human sociality. His pigeons help him to win *khokhā* competitions where both economic capital (money) and symbolic capital (*izzat*, prestige, and status) are at stake. Although the role of money remains secondary in *khokhā* competitions, the focus is on achieving honour. The loser is humiliated and mocked whereas the winner asserts himself for having superior skills. To a South Punjabi pigeon flyer, the most prestigious reward is when his friends and opponents acknowledge his mastery, praise his training techniques, and appreciate the colour and beauty of his birds. Such status elevation is mostly experienced by *ustāds* (pigeon masters) who are revered by their loyal disciples for their years of knowledge and experience.

Pigeon masters—the centre of authority and respect in the *Asmānī* competition

Ustad Shakoor was in his late 60s when I first met him in 2014. "I was born a year before the creation of Pakistan (1947)" he replied when I asked about his age. A short, lean man with a short beard, he was known as a seasoned pigeon enthusiast among the flyers. He worked all his life as a milkman, carrying milk-cans from different villages to the town of Kahror Pacca on his bicycle, and this physical activity contributed to keeping him fit. At the time of Partition, Shakoor's father used his extensive social capital to acquire two shops in the main bazaar that belonged to a migrating Hindu merchant. He later passed on these shops to his two sons: Shakoor and Qasim. Shakoor always saw this shop as his financial security, and thought he would sell it one day to support his only son's higher education, or to bribe a politician to give his son a government job—the dream of every South Punjabi father. However, soon as the son turned sixteen, he asked Shakoor to sell the shop and give him money to set up a tea stall just opposite the main bus station. Shakoor rebuffed the "absurd" proposal of his teenage son, and the outraged boy swallowed a wheat pill in anger (the pill contains aluminium phosphide, used as a fumigant for stored wheat grain).

"Nobody knows how Ustad Shakoor coped with the 24-hours-long memory of watching his only son die in misery, but he started spending more and more time with his pigeons," Rafik told me while explaining his *ustād*'s story. After that incident, Shakoor devoted himself completely to his pigeons and started to achieve mastery in breeding, training, feeding, and healing the bird. He invented some secret formulas by mixing an exact number of different herbs and seeds to prepare a *goli*, or a pill, that was an all-rounder: it could increase the pigeons' stamina and make them fly in rain, enhance their attention span, and even make them forget their thirst in the unbearable heat of June. Witnessing his mastery, Rafik and many other flyers submitted themselves as dedicated disciples (*shāgird*) to the *ustād*. Many of them visited him every day to learn from him and, as one of his disciples told me, "to absorb some part of his immense knowledge of pigeons." When Shakoor fell ill with malaria in 2015, one of his disciples bore the expense of his medical treatment. Other disciples contributed by purchasing fruits, vegetables, and packets of wheat flour for the master's household until he recovered. Such a relationship between a master and disciples is common in South Asia where it is an implicit expectation that in return for the knowledge given by the master, disciples reciprocate with care and material goods in times of need (see Frembgen and Rollier 2014, 18; Alter 1992, 59).[9]

This relationship of master–apprentice has been documented extensively, especially in societies where personalised schooling is considered important.[10] Among South Punjabi pigeon flyers, the relationship between master and disciple is personal, since it is a result of sharing the same *shauq*. To maintain this relationship, all *ustāds* strive hard to ensure the success of their disciples by providing them secret instructions on the pigeon's dietary and training techniques, and by helping them during pigeon flying competitions. However, there is also an implicit lament about changing times and how such relationships are increasingly compromised under the onslaught of modernity. A new generation of *ḍarī* pigeon flyers (usually known as *chiṛī mār*, see Chapter 6) seems less interested in nurturing this relationship of respect and obedience to masters. Some young flyers hold that the time of true *ustāds* is now over and if there are some good ones still around, finding them is very difficult. Others argue that *ustāds* tend to keep their knowledge to themselves and rarely share everything with their disciples, as Khudla recounted, "I offered a cow to my *ustād* in exchange for his complete *ustādī* (mastery), but he refused." A similar claim was made by some *khokhā* pigeon flyers such as Aslam who sneered when speaking of his late *ustād*, "He never shared his wisdom with any of his disciples and now he is dead. Who will benefit from his secret knowledge?" I also met a few young *khokhā* pigeon flyers who, when asked about their master, responded curtly: "we are our own masters." The decline of the master and disciple tradition in present-day South Punjab, as I argue later in the book, may be seen as part of a wider transformation of Punjabi society with a fast-paced lifestyle and a changing value system that places a premium on material goods and consumption rather than on traditional practices built on culturally valued principles and social relations.

The importance of *ustāds* may seem to be declining among *ḍarī* and *khokhā* pigeon flyers, yet many *asmānī* pigeon flyers continue to honour this tradition. Some flyers tie a white turban (a symbol of respect) around their *ustād's* head to formally announce him as their "master for life" (*zindgī bhar dā ustād*). Others begin a pigeon competition with the blessings of their master. In my discussion on *asmānī* pigeon flying competition below, I show why and how the tradition of maintaining utmost respect and admiration for the *ustad* has survived as a central part of this activity despite changing times.

A very different breed of pigeon is used in the *asmānī* competition. Unlike *golay* pigeons who fly in the shape of a flock and are used for *ḍarī* and *khokhā* competitions, *asmānī* pigeons (or tippler) are famous for their endurance and can fly high up and sometimes remain out of sight for fourteen to sixteen hours at a stretch.[11] The ability of these pigeons to reach such heights in a spiral ascent and remain there for an extended period renders them an expensive entity, and a good quality *asmānī* pigeon is sometimes valued at $3000. This competition is a popular feature in the cities of central Punjab, such as Lahore, Sialkot, Kasur, and Faisalabad. Anthropologists Jürgen Wasim Frembgen and Paul Rollier (2014, 58) in their study of three popular pastimes—wrestling, pigeon fancying, and kite flying—estimated that more than 100,000 people of all classes are actively involved in the practice of keeping and flying pigeons in Lahore. The authors devoted a chapter to pigeon flying in their book, examining the sacred status of the bird in Islam, the performative element of pigeon flying, the effect of this activity on the lives of the practitioners, and the relationship between a master and a disciple. In Lahore, they suggest, the annual *asmānī* pigeon flying competitions are advertised through posters, involve an entry fee, and offer tangible prizes, including money (sometimes reaching up to 400,000 Rupees or $2500), television sets, fans, motorbikes, as well as trophies (Frembgen and Rollier 2014, 69–72). Other than pecuniary benefits, the main purpose of keeping and flying pigeons in Lahore (as in South Punjab) is to win *izzat* among peers and to fulfil personal *shauq*.

Almost 500 kilometres away from Lahore in South Punjab, the *asmānī* competition has quite a different aura. The pigeon flyers here can only afford a few low-priced *asmānī* pigeons that cost between $6 and $30. The competition is far less commercialised and takes place in the early mornings of the hottest summer months, mostly from May to July. The seasonal importance of the competition is crucial, since in winter *asmānī* pigeons fly away for days on end and often fall prey to predators. "There is no greater pain than losing a pigeon to a hawk," Ustad Jameel told me. The summer heat, according to him, works in the favour of flyers by making the bird land for water before nightfall.

A clean-shaven old-timer in his late 60s with a benign appearance, Ustad Jameel was often surrounded by his disciples whom he helped in preparing special concoctions to enhance their bird's performance. However, his role was more than that of a mentor, and he was usually active in planning *asmānī* competitions and helped participants agree on terms and conditions of the event including the time, date, rules, and amount of betting money. This is the only competition where betting is explicitly involved, and masters ensure that the bets are lodged fairly. For this purpose, Ustad Jameel usually collects $20–$30 from each participant a week prior to the competition and marks each pigeon with a stamp. As the birds are released in the early morning, he closely observes any foul play, including whistles or shouting, which are strictly prohibited in this competition. He stays on the rooftops the entire day, keeping himself busy by sipping hot tea and observing the birds through binoculars despite the uncomfortable heat of the midday sun. In the late afternoon, as the pigeons begin to return from their aerial tour, he carefully calculates their flight time. As long as the pigeons land on their roof before sunset, he adds up their time and announces a winner with the longest flight time.

To many pigeon flyers, Ustad Jameel is not only an embodiment of knowledge and wisdom, he is also a true *shauqeen*. Ustad Jameel told me that organising an *asmānī* competition or staying on the rooftop under the blazing sun is "only possible when you have a genuine *shauq*." The competition is a meaningful activity to him, in which he enjoys each minute and fulfils his enthusiasm by engaging with other enthusiasts. His disciples and other *asmānī* pigeon flyers of the area recognise his contribution and admire him for his commitment and dedication. I met four other masters during my fieldwork who organised and managed *asmānī* pigeon flying competitions from start to end, and received great respect from their devoted disciples. As they regulated bets, I never saw their authority and fairness being questioned.

The wagering amounts in the *asmānī* competition are relatively small since the actual competition is over symbolic gains (*izzat*). To ensure one is well prepared to compete for his *izzat*, pigeon flyers work tirelessly on their birds for days before the competition, feeding them special diets and training them regularly. At the time of competition, when the money is wagered in front of other flyers and *ustads*, it stands symbolically similar to wagering one's *izzat*, pride, and skills. As pigeons fly in the air, they not only become an extension of their flyer's personality and an embodiment of his hard work, passion, and commitment, they also become surrogates of their keeper's honour and actively influence his *izzat* among peers. Thus, the allure of flying pigeons is more than just amusement or financial gains; the men fly their birds as a way of enhancing their prestige among peers.

However, the struggle for symbolic gains (honour and status) does not mean that money is not important. Many South Punjabi areas are poverty-stricken, where most Seraiki-speaking people experience structural inequalities, deprivation, rooted illiteracy, and unemployment. Many people earn less than $4 a day and this has led to a resentment of the governance of the province that has largely been dominated by central Punjab. To seek alternative avenues for livelihood, many South Punjabis migrate to Lahore, Karachi, and Rawalpindi, and sometimes to Saudi Arabia, Qatar, and United Arab Emirates to work as unskilled labour. Although pigeon flyers avoid migration to remain close to their birds, what little they earn is spent mostly on their cherished birds. They hope that pigeon competitions will bring some financial reward to compensate for these high costs. However, according to my calculation, it is rare that monetary gains match the expense on the diet and medication of the pigeons. The flyers I spoke to also acknowledged this and many rationalised material rewards as peripheral to this activity. As Khulda put it: "A man can sleep on an empty stomach but he cannot sleep with shame." For Khudla, Rafik, Aslam, Ustad Shakoor, Saifal, Chacha Munda, and many others, pigeons were not a means of livelihood, rather, they were cherished companions with whom the men can fulfil their *shauq* and demonstrate their skills and expertise to their peers to achieve honour.

Although success in pigeon competitions does not alter a flyer's economic standing, it does enhance his status among his peers. This is very much in line with Bourdieu's (1986) observations about the efficacy of forms of capital within different fields. Each field, he maintains, has its distinct rules through which individuals are evaluated and ascribed social position. In other words, it seems that when flying pigeons, the men establish a form of capital aimed at accumulating *izzat* and other symbolic rewards by participating in economic transactions through betting and monetary rewards. The birds can, therefore, be viewed as both currency and symbolic expression of their honour, prestige, and respect. This approach enables us to view pigeon flying as a social, economic, and symbolic practice that exceeds the act of playfulness. However, this does not mean that pigeons are passive vehicles for their keepers. Rather, as we have seen, each pigeon embodies the keeper's flying style and is a testament to his emotional, economic, and physical investment.

Earlier, I have suggested that a description of the *asmānī* pigeon flying competition will illustrate how and why the tradition of masters and disciples has survived despite changing times. Above, I have showed that a master's role in organising the *asmānī* competition and managing the bets is indispensable. The presence of a respected master also eliminates any chance of cheating and ensures fair play. Second, as the structure of the *asmānī* competition is different from *ḍarī* and *khokhā* competitions, and chances of a flyer's success depend on the pigeon's skill, thus the knowledge and advice of *ustāds* in making a feed that enhances the pigeon's stamina is highly valuable. Third, a different breed of pigeons is used in *asmānī* competition and requires careful breeding to preserve the bloodline. Therefore, many flyers seek the advice of their masters who, because of decades of experience and close relationship with pigeons, are believed to be experts in selecting the best mating pair.

Fourth, *asmānī* pigeons are more expensive as compared to *golay* pigeons and thus, for a utilitarian reason, more effort is devoted to their health and safety. Pigeon masters, therefore, prove critical in providing their expert medical advice to cure sick birds. Lastly, although the rules of *asmānī* pigeon competition are widely understood by all flyers, it is the "tradition" (*riwāyat*) that an *ustād* announces the winner at the end of the day. When a flyer receives admiration from an *ustād*, it raises his joy in success. Overall, these and other related reasons are instrumental in maintaining the sanctity of the centuries-old master–disciple institution among rural South Punjabi pigeon flyers, and this ensures that the wisdom of *ustāds* is respected despite changing times.

The captivating enthusiasm

Through the discussion of the three types of pigeon flying competitions in rural Pakistan (*ḍarī*, *khokhā*, and *asmānī*), and an examination of the breeding, feeding, training, and decorating practices of pigeons, I have argued that pigeon flying is a meaningful activity for men in South Punjab that not only provides them with an avenue to earn important symbolic rewards (*izzat* and prestige), but also enables them to achieve a sense of fulfilment, freedom, and true pleasure despite serious life problems. Pigeons, in the words of Donna Haraway (2016, 16), "are competent agents—in the double sense of both delegates and actors—who render each other and human beings capable of situated social, ecological, behavioral, and cognitive practices." To pigeon flyers like Qamoo, the coachman who saw a sudden decline in his business, or Ustad Shakoor, the milkman who witnessed his only son commit suicide, pigeons enabled them to overcame their hard times by building on profound cross-species affection. To other pigeon flyers like Rafik, Aslam, and Khudla, pigeons made it useful to counter difficult circumstances at various stages of their life. Through such stories, I show that the company of pigeons and their physical presence in the household can transform this seemingly mundane activity into a captivating enthusiasm.

An important ingredient in developing intimacy with pigeons lies in the bird's loyalty, which is usually gauged in three ways. Firstly, it is the loyalty of the bird that helps the flyer take his pastime as a serious pursuit. The more-than-human relationship developed in the company of pigeons cultivates a way of life in which enthusiasts achieve a sense of continuity with the *shauq* of their fathers, uncles, and grandfathers. Therefore, the loyalty of pigeons enables the men to keep this tradition alive by passing on this generational enthusiasm (*khāndānī shauq*) to subsequent generations. Secondly, the pigeon is depicted as a loyal bird in mainstream Islamic discourse and iconography and many flyers take this to support their attachment to the bird.[12] Rafik and others often recount a story in which pigeons played a pivotal role in saving the life of Prophet Muhammad. The story goes that some non-believers followed the Prophet with murderous intent when he was migrating from Mecca to Medina in 621 CE. He took refuge in the cave of Thawr and a spider immediately spun a large web at the entrance of the cave while pigeons made a nest nearby. Upon arriving at the cave's entrance, the enemies found the pigeons nesting peacefully while the spiders' web was unbroken, so they continued their pursuit in the opposite direction. This story, according to pigeon flyers, is a testimony to the devotion of pigeons and has since made Muslims recognise the bird's loyalty to Islam (see also Jerolmack 2013). Some men believe that the story alludes to how keeping pigeons is a spiritual quest rather than a frivolous hobby. Lastly, as pigeons pair for life, they are seen as an epitome of devotion and fidelity. Pigeon flyers spoke passionately about how both male and female pigeons brood over the eggs and collectively nurse chicks with pigeon milk. Many flyers appreciate this quality in pigeons and believe that the bird's faithful and affectionate disposition extends to their human caretakers.

The loyal demeanour of pigeons reflected in their bodies and apocryphal stories make rural South Punjabi flyers feel proud of their *shauq*. Many men continue to take this *shauq* to develop an intimacy with their cherished birds and form an understanding of their preferences through careful domestication and close association. Such close attachment is also evident in the *shauq* of cockfighting, where enthusiastic cockfighters carefully feed and train their roosters and expect the bird to fight with strength and courage in the arena to help them win masculine honour among peers.

Notes

1 More than 100,000 pigeons were used in WWI, and some of them won medals of honour. One of them was Cher Ami (French for "dear friend") who won the Croix de Guerre with Palm for saving the lives of 194 soldiers by delivering their distress message to base camp with great difficulty (Trueman 2015).

2 In Lahore, I learned that pigeon flying is sometimes described as sport (personal correspondence with Jürgen Wasim Frembgen). However, in the area of my research, people seldom use the word sport for these activities. For more on the use of the word "sport" for traditional activities, see Besnier and Brownell (2012, 446–447).

3 Among New York's pigeon racers, Colin Jerolmack (2013, 159–160) notes, such a pigeon is called "chico."

4 A practice in which horseback players compete to carry the carcass of a goat or a sheep to a designated place to score a point.

5 Evans Pritchard's (1940, 41–46) famous study among the Nuer and Fijn's (2011, 95) ethnographic observations among Mongolian herders also show how animal categorisation through their colours and patterns is a crucial part of human–animal sociality.

6 Some flyers speculate that *shirāzī* pigeons were imported from Shiraz, Iran. The word *shirāzī* literally means "from Shiraz." In *Ain-i-Akbari*, Abul Fazl suggests that Akbar kept this breed specifically for the beauty of its colour (1873, 302).

7 Such love of red pigeons resonates with the larger Islamic traditions concerning the bird. When I entered the library of Madrassa Qasim ul-Uloom in Multan in a quest to explore Islamic accounts on domesticating pigeons, I found a chapter

on *hamama* (Arabic for pigeons) in Al-Dameeri's famous book *Hayat-ul-Haiwan* (The Life of Animals). I was not surprised to read a tradition about how Prophet Muhammad liked to watch red pigeons (2006, 589).

8 See for example, Alter (1992, 120), Frembgen and Rollier (2014, 54), Shaw (2000, 200–201).

9 Frembgen and Rollier (2014) note that in Lahore and other areas of the Punjab, there is an additional figure to the *ustāds*: the *khalīfā* (literally, a successor). In some areas, *khalīfā* is a senior apprentice who can become *ustād* with experience. In other places, the *khalīfā* is more respected than the *ustād*.

10 See Kippen (1988, 133), Alter (1992, 60), Frembgen and Rollier (2014, 18), Herzfeld (2004, 67), and Marchand (2014).

11 In addition to *golay* and *asmānī* pigeons, homing (or racing) pigeons are known for their speed and ability in finding the way back home across great distances. Mostly used in Europe, these racing pigeons win huge prizes and have been sold at auctions for millions of dollars. There are no racing pigeons in South Punjab and few enthusiasts keep *asmānī*, while most pigeon flyers cherish the company of *golay* pigeons.

12 Many flyers believed that pigeons carry beatitude because of their association with shrines and mosques. Some even argued that pigeons' sound "hu" is similar to Allah Hu, a Sufi chant in remembrance of God (*zikr*). Such beliefs can be traced to the accounts of the Prophet who advised his cousin and son-in-law, Ali, to keep pigeons in his house and when the pigeons start cooing, he should start *zikr* or worship (Al-Dameeri 2006, 589).

3 The seduction of cockfighting

Forbidden dangers

"If Prime Minister Nawaz Sharif were to personally ask me for my rooster, I would refuse," Malik Khizar Bohar remarked while showing me his prized hen and gamecock. A middle-aged man of average height with black eyes and a short, salt-and-pepper beard, he wore well-polished Peshawari sandals and a neatly ironed kurta shalwar. Such careful choices reflected his personal taste and also suggested his noble (*ashrāf*) background. Like many other landlords of the area, he inherited 100 acres of agricultural land from his father but has sold almost half of it over the years. Mostly, this was because of his passion for cockfighting that involved large sums of money spent on purchasing, training, fighting, and gambling on roosters. When he came to know that I had come all the way from Australia to "research" cockfighting, he sent his car and chauffeur to fetch me to his place where a table laid out with a heavy breakfast welcomed me. There in his house, as Malik Khizar filled up my glass with Coca-Cola, he started describing his *shauq* of cockfighting.

"Gamecocks are the bravest creatures on earth; they fight until the last breath," he stated confidently. "Have you ever seen a bird taking a hit on his face and still flying?" He continued, "This is the only bird that fights with kicks, sometimes receiving 200 blows to the face, but he still keeps going." It all began about fifteen years ago when a friend took him to a cockfight, he told me. He watched this display of courage for the first time in his life, and immediately "fell in love" with this bird. For years, he learned the finest details of breeding, feeding, and training gamecocks, and developed friendships with famous cockfighters in the Punjab and in Sindh provinces. The time, attention, energy, and resources spent on his *shauq*, he said, was a way to achieve "personal comfort" (*zāti sakon*) and to accomplish his *tharuk* (a kind of deviant attraction).

The *shauq* of cockfighting, according to Malik Khizar and many other cockfighters of the area, is accompanied by *tharuk*, which involves a series of acts including gambling, participating in cockfighting organised at remote and sometimes secret places, and mocking and humiliating the loser after the fight. Both *shauq* and *tharuk* allow *shauqeen* cockfighters to appreciate what are seen as paramount masculine qualities of gamecocks, including fighting with courage (*jur'at*), staying fearless and carrying on with bravery (*dilerī*), possessing the strength to dominate the opponent (*zor*), and not surrendering even when in a difficult situation. It is because of this thrill and frenzy that Malik Khizar and many other cockfighters of the region cherish the company of their birds and of cockfighting friends over their family, neighbours, and relatives, and treat their personal passion as a precious enthusiasm.

Although Malik Khizar was a wealthy man, in rural South Punjab cock-fighting was not limited to the elite but attracted people from diverse socio-economic backgrounds such as landlords, village notables, government officials, village school teachers, police officers, small farmers, shopkeepers, bus-stand managers, fruit vendors, and daily wage labourers. There were at least 200 cockfighting enthusiasts in my field site whose strong attachment to their birds shaped their social life. Some men were recognised as expert breeders, others as champion trainers, and some were known only as enthusiasts. Yet, all *shauqeen* treated each other with respect despite class, ethnic, and sectarian differences, and only those were ascribed a superior position within the group who had excelled in the field, and gained *izzat* through their mastery in and dedication to this *shauq*.

By engaging in the everyday experiences of cockfighters and examining their sociality with their gamecocks, I focus on multiple social and economic

Figure 3.1 Malik Khizar (left) showing a pair of precious breed chickens. His helper (right)—who was not a cockfighting enthusiast—accompanied him.

arrangements that drive and frame this *shauq*. In discussing the impact of cockfighting on shaping relationships between humans and roosters, I argue that this *shauq* is not only a source of personal joy but is also a meaningful activity to achieve honour by establishing hyper-masculine traits among peers. I contend that cockfighting as a *shauq* enables us to explore the values that underpin interpersonal relationships in rural Pakistan and how such relationships are maintained and transformed through active participation of non-human lives. Cockfighting thus emerges as "serious leisure" (Stebbins 2015), demanding a strong commitment of time, energy, and resources for developing a close intimacy with a rooster during the breeding, feeding, and training process. This close human–cock relationship is tested in the cock-fighting arena where dominant traits of rural Pakistani maleness, such as courage (*jur'at*), bravery (*dilerī*), and strength (*zor*) are performed, reproduced, and reaffirmed through the fighting birds. For this reason, instead of viewing the passion of cockfighting as a futile activity, I explore it as a cultural force that animates the *shauqeen's* social universe and enables them to win *izzat* (honour), perform *mardāngī* (manliness), transcend status, and cultivate life-long friendships.

My interest in studying the culture surrounding cockfighting began in 2008 after reading Clifford Geertz's seminal essay on the Balinese cockfight. This essay, first published in 1972 and later included in his well-cited book *The Interpretation of Culture* (1973), emerges as an ethnographic account to support Geertz's proposed paradigm of interpreting cultures through thick description. With his evocative description and critical insight, Geertz views the cockfight as an expressive social event, a cultural text that allows a "Balinese reading of Balinese experience, a story they tell them-selves about themselves" (1973, 448). He centres his analysis of cockfights on two forms of betting: deep fights that involve high stakes and take place between "the solid citizenry around whom local life revolves," and shallow fights that involve smaller amounts of betting and are favoured by the poor (1973, 435). According to Geertz, deep fights hold special significance for Balinese people because through these "the really substantial members of the community" vie for status (1973, 435). However, this status gambling, as Geertz points out, does not affect the social hierarchy in Balinese society because from the outset the high stakes exclude poor gamblers.

Even though the loser is humiliated and the winner is applauded in the cockpit, Geertz stresses, "no one's status is actually altered by the outcome of a cockfight; it is only, and that momentarily, affirmed or insulted" (1973, 433). He further continues:

> The cockfight is "really real" only to the cocks—it does not kill any one, castrate anyone, reduce anyone to animal status, alter the hierarchical relations among people, or refashion the hierarchy; it does not even redis-tribute income in any significant way.
>
> (1973, 443)

Geertz's powerful description of cockfighting, despite appearing about half-a-century ago, is instructive and assisted me in exploring the cultural meaning of cockfighting (along with pigeon flying and dogfighting) in South Punjab. Yet, critics are right in pointing to the key weakness in his text: his overemphasis on men's world and the fact that an activity cannot be regarded as "cultural text" when it privileges male interpretation (Roseberry 1982, 1019–1020). Also, Geertz's description lacks any analysis of the historical development of the Balinese cockfight (see Howe 2005, 31), and leaves us questioning the transformation of the human–animal and, more specifically, the human–rooster relationship in Bali. As studies examining the bullfight (Marvin 1988), horse racing (Cassidy 2002), elephant encounters (Locke and Buckingham 2016), or the cow protection movement (Govindrajan 2018; see also Narayanan 2018) have demonstrated that if the human–animal relationship is observed through a sociohistorical lens, this can provide us with important details about the social structure and everyday cultural patterns. Moreover, according to some critics, Geertz's tends to rely on his own subjective interpretation rather than engaging in dialogue with the Balinese who "remain cardboard figures" (Crapanzano 1986, 71). Hence, Geertz's interpretive approach comes at a cost, where claims are asserted and investigated but "without providing data that substantiates them" (Jacobson 1991, 53).

My purpose in this chapter is not to address such criticism through my description of South Punjabi cockfighting but, while keeping these in mind, I take inspiration from Geertz's analysis to construct the social world of rural Pakistani cockfighters. However, in doing so, I heed Paul Stoller's (1997) advice and take my analysis beyond the visual to include the sensual. In his *Sensuous Scholarship* (1997), Stoller suggests that an analysis must go beyond the sense of vision and incorporate "smells, tastes, textures and pains" (1997, xiv). He convinces scholars to "track between the analytical and the sensible," and understand and interpret embodied engagements in a sensuous way (1997, xv). Accordingly, I contend that the human–cock relationship in South Punjab cannot be analysed only as cultural text but that interpretation requires a deep understanding of the interplay between different senses—the sound and smell of cocks, the taste of their feed, the tactile nature of training, and the sight of the cockfight. Such an examination will help in understanding this activity holistically, taking into account the intimacy developed between a man and a gamecock through activities of breeding, caring, feeding, training, and fighting the bird. A sensuous understanding of this activity mixed with a Geertzian interpretive analysis will, in my view, enable us to explore why people indulge in cockfighting, structure their lives and daily routines around their roosters, and expect crucial symbolic rewards from this activity. Such an analysis will lead us to explain and understand the human world in conjunction with the animal world, and consider how this inter-species relationship between a man and his bird develops into emotional attachment between them that is tested during the climactic fight.

The South Punjabi practice of cockfighting (locally known as *kukaṛ-bāzī* or *murgh-bāzī*)[1] demands a complete refashioning of the self and deeply influences the men's social life. In what follows, I employ a sensuous analysis to examine the human–cock relationship by explaining men's affection towards their roosters that starts before the hatching of the egg and takes different forms over the bird's lifespan. I then discuss two forms of cockfights—"bare-heel" cockfight and "covered-heel" cockfight—to emphasise their distinct structure and organisation. I then explore how betting coexists with *shauq* and how it structures relationships between friends. In rendering cockfighting as a series of activities and ideologies rather than as an isolated event, I show throughout the chapter that this *shauq* demands that enthusiasts take their roosters as subjects and not as passive objects, thus intensifying their passion by active engagement with the birds.

Men and gamecocks: a sensuous inter-species relationship

My first thought on roosters goes back to my childhood when the primary school teacher ordered that I become a *murghā* (or a rooster). This was a punishment administered to those untamed students who did not respond to being caned by the teacher. Bent double with arms going through the knees to hold the ears, or at least touching them, was not an easy position and after only two minutes, sharp pains ran through the arms and legs, flushing the face and making the head swing. To an adult, this might seem a most humiliating position, but to us third-grade school children, becoming a *murghā* was a part of school routine.

The rooster was also present during my childhood, and at the time of my fieldwork, as the local version of an alarm clock when cockcrow at dawn meant a chorus of village roosters competing with each other in making the loudest noise. This is how most peaceful summer night slumber would end. Later, in a suburb in Canberra, when a neighbour brought an energetic rooster to add to his backyard chicken coop, memories of village life came back to me; of nights spent under the stars lying on the wooden charpoy, the sublime sunrise, and the distinctive call of koels. It also brought back memories of childhood, running wild in the wide courtyard, playing cricket, and breakfast with lassi (a drink made with yogurt). The only thing missing in Canberra was the sound of cockcrow mixing with the loud call for prayer, or *āzān*, recited at dawn by devout Muslims through the mosque's loudspeaker to wake up people for prayer.

The rooster's unflinching ability and commitment to crow exactly when it was time for the dawn prayer made some people consider him a "holy" bird. In Seraiki language, the word for cockcrow—*bāng* or *āzān*—is also commonly used to describe the five daily calls to prayer for Muslims.[2] This ability of roosters to rouse Muslims for the dawn prayer was appreciated by the Prophet of Islam, and according to some sources, he forbade Muslims from reviling the bird (Abu-Dawud 18, 220) because roosters crow when they see an angel

(Al-Bukhari 59, 111). Thus, the presence of a rooster in a house is commonly regarded as a source of blessing from God, and as Roberto Tottoli (1999) argues, this means the bird is held in high esteem in the entire Muslim world.

South Punjabi cockfighters like Waso feel a special pride in their association with this "sacred" bird. Waso, a tall man in his early 50s who, with his thick-soled boots and grey turban seemed even taller, lived with his wife in a large house near Kahror Pacca town. When I first met him, his full-grown gamecock was roaming idly in the courtyard followed by two hens, only sometimes pecking at the earth to catch a worm. Waso recounted his life-long enthusiasm for keeping and fighting cocks, why he liked this activity, and how people gave him *izzat* because of his *shauq*. After about ten minutes, as I sipped the sweet creamy milk tea, the *dodh pattī* prepared by his wife, I asked Waso if he knew the Islamic position on fighting this "holy" bird. "You need to understand the difference," he replied, "There are different roosters, and some are specifically created by God to fight." I was surprised. "Look at this *aṣīl* rooster" Waso pointed towards his gamecock who was now standing on one leg, "If you leave him in the jungle, he will fight with other roosters. It is in his nature." He was convinced that this distinction is important, and for this reason he never eats his fighting cocks. "The one I eat (farm chicken) is different from the one I love (a gamecock) ... fighting cocks taste worse than a broiler." He argued that daily exercise and massage make the gamecocks body hard "like stone" and if cooked they do not taste good. However, he suggested that gamecock meat is a male aphrodisiac because the bird consumes a good number of almonds, pistachio, and a high protein diet. For this reason, traditional healers (*hakīm*) often recommend *desī* cocks to be cooked in *desī ghī* to increase general physical strength and to restore vitality.

Aṣīl chicken, the one that Waso indicated are created by God to fight, is the most popular breed in rural Punjabi cockfighting. The word *aṣīl* comes from *aṣlī* or *aṣal,* meaning "original" or "authentic" and reinstates the local belief that this breed is still in its original form—unsullied by the introduction of cross-breeding by humans—and is, therefore, the most authentic fighting breed in the world. This breed is different from farm chickens and even from *rezā*—the second preference among cockfighters, famous for its speed and agility but shorter in height and with less stamina compared to *aṣīl*. *Aṣīl*, many believe, is all about power, stamina, and stature—a truly masculine gamecock.

Both *aṣīl* and *rezā* come in different colours such as *jāwā* (off white plumage and black body), *lākhā* (dark brown and shiny green plumage and black body), *kāle mushkī* (complete black), *chiṭṭe* (complete white), *pele* (yellow neck and tail with brown plumage), *lāl* (rusty-red), *kāsnā* (light-grey), and *sāwā* (shiny green). Like racing horses in England (Cassidy 2002, 91) or domesticated animals in Mongolia (Fijn 2011, 101), distinct colour fighting cocks are given personalised names to draw them into the human domain and reflect their individualised traits including their fighting character, history, and distinct attributes and abilities (Affergan 1994, 202–204; see also Marvin 2005). Some famous names such as Mithā (sweet), Kābrā (cobra),

Thānedar (police constable), Dilbar (close to the heart), Suhrāb (hero), Jānbāz (fearless adventurer) reflect the rooster's skills, fighting style, and behavioural pattern. For instance, Kābrā rooster is famous for its attacking skills, much like a cobra, and its ability to raise its neck in the course of a fight. Thānedar possesses a rowdy personality, much like a police officer; Suhrāb's heroic character renders him a winning cock; Jānbāz is seen by his keeper as fearless; while Mithā and Dilbar are cherished by their keepers for their exceptional physical traits. These unique names make the gamecocks a distinct being, differentiate them from village chickens, and allowed the enthusiasts to praise and recognise their individual personality and qualities.

In the context of South Punjab, pedigree is important and it is crucial for a rooster to belong to a good breed (*nasal*). "A good breed is the key to success in cockfighting," Ustad Ramzan Mochi informed me in the early days of my fieldwork. A cobbler by profession, Ustad Ramzan was a retired cockfighter, a worn man in his 70s with a benign appearance. From the 1970s to the 1990s, he developed an interest in keeping and fighting partridges and participated in competitions all over Pakistan, winning over two dozen trophies. He was given the tile of *ustād* (master) by his disciples. However, in the last decade of the twentieth century, interest in partridge fighting faded in South Punjab, so he shifted to cockfighting, "I joined cockfighting because of my pedigreed hen (*naslī murghī*). When she died, I left cockfighting." Widely popular as an avian *hakīm* (a traditional healer), Ustad Ramzan now provides advice to cockfighters, one of them being Imran Niazi.

Imran Niazi, referred to as Mani by his friends, was in his early 30s and worked as a bus-stand manager when I first met him. His father, a police constable and a famous cockfighter of his time, migrated from Mianwali district to Kahror Pacca in the 1980s. Mani was born and raised in Kahror Pacca, where

Figure 3.2 Aṣīl pelā (left) and *aṣīl lākhā* (right).

he later met Ustad Ramzan and became a loyal disciple. Unlike pigeon flying, there is no systematic and longstanding tradition of the master–disciple relationship in South Punjabi cockfighting. This is perhaps because every cockfighting enthusiast keeps a small number of roosters and uses his own mastery (*ustādī*) to breed, train, and ready the bird to fight well. Nevertheless, Mani showed great respect for Ustad Ramzan and followed the instructions of his frail master diligently. When I first met Ustad Ramzan in January 2015, he called Mani and we sat together cross-legged on a carpet at Ustad Ramzan's place. Amused by my interest, they started explaining different feeding and training methods for gamecocks, however, their focus was on breeds. "*Aṣīl* is an honourable (*ghertī*) chicken, a thousand times better than *rezā*," remarked Ustad Ramzan. "Yes, *rezā* cocks are very quick in attacking, but they cannot bear a single blow," corroborated Mani. Though *aṣīl* was a favourite choice for cockfighting, they both suggested that not all *aṣīl* displayed equal courage in a fight.

Such courage came, according to Ustad Ramzan, through the "mother" rather than the "father." Thus, during breeding, he suggested you need to carefully select a hen that has a history of producing fearsome cocks. He averred that in two-legged animals, the breed is carried forward by the mother, while four-legged animals generally possess the qualities of their father. However, Mani remarked, this does not mean that the role of a cock is insignificant in mating; "a champion's son will become a champion." Concurring with his master about the vital role of the hen, he added: "a gamecock from a *naslī* (good breed) hen can die fighting but will never flee from a fight." A fleeing (coward) cock, as we will see below, is usually labelled as *bad-naslah* (mixed breed or bad blood).

In South Punjab, the hen is not selected because of her "beauty" or on her external characteristics, as is the case in the Philippines (Guggenheim 1994, 144), rather, she is selected for her pedigree and some key physical features. For instance, a good hen is believed to possess a white beak, white eyes, white shanks, short neck, small neck bone, medium-sized cheeks and jaws, and a small comb. When a hen with most of these physical features is available, cockfighters try to mate her with a cock who has a shine in his plumage like a snake, a long tail like a parrot, a long-shaped mouth like a mongoose, folded wings, big thighs, small shanks, and a sharp beak. Such embodied knowledge about roosters shows the depth of a cockfighter's *shauq*, and indicates his close sociality with his chickens developed through the continuous investment of time. Time thus becomes a capital, something that when invested with care develops more-than-human sociality and results in success in the cockfight arena. In this sense, time can be perceived as a kind of symbolic investment, in line with Bourdieu's observations, that "time must be invested, for the value of symbolic labour cannot be defined without reference to the time devoted to it" (1977, 180). By spending time in breeding, and most importantly in getting to know the cock's body and behaviour, the *shauqeen* create a bond that assures them of the rooster's capabilities and prowess. In Bali too, Geertz emphasised, cockfighting enthusiasts invest hours in training and feeding their gamecocks to prepare them for the fight:

> A man who has a passion for cocks, an enthusiast in the literal sense of the term, can spend most of his life with them, and even those, the overwhelming majority, whose passion though intense has not entirely run away with them, can and do spend what seems not only to an outsider, but also to themselves, an inordinate amount of time with them. "I am cock crazy," my landlord, a quite ordinary *afficionado* by Balinese standards, used to moan as he went to move another cage, give another bath, or conduct another feeding. "We're all cock crazy."
>
> (1973, 419)

Because grooming a champion cock in South Punjab requires years, this sacrifice of time becomes something personal—a cherished investment that keeps cockfighters from sharing their bird with others for breeding. Many of them would be reluctant to breed a bird with the opponent even if a large sum of money were offered to them or if someone with a particular political or social authority demanded this. Ustad Ramzan recounted the fable of a poor farmer who did not give his pedigree birds to the Nawab (ruler) of the State of Bahawalpur. The story goes:

> The Nawab once visited a farmer's house to get his superior breed hen and cock. The cock was 20-inches high with shiny feathers and long spurs and never lost a single cockfight. The hen was also tall and beautiful with a shiny red comb. Upon the Nawab's arrival, the poor man noticed his hidden intent and asked whether he could cook something in honour of his visit. When the Nawab agreed, the farmer slaughtered his treasured chickens and asked his wife to cook them well. When the Nawab finished eating, he told the poor man the purpose of his visit and asked for the farmer's pedigree hen and cock. The poor farmer smiled and replied, "Nawab sahib, you have just eaten them."

This fable illustrates several points: the ethos of hospitality, hierarchy, and subordination of the rural poor. It also highlights the politics of taste that enabled the farmer to exercise his agency in choosing to surrender his chickens to the Nawab by cooking them instead of being forced to part with his prized possession. The poor man was clever enough to deploy the cultural expectations associated with hospitality to his advantage, and presented his chickens to the Nawab as a form of gift that ultimately undermined the latter's authority. Arguably, the cock and hen were the inalienable possessions of the farmer because they were bred by him, trained by him, and invested with his affection. Even when consumed by the Nawab, the birds remained a gift attached to the farmer and an embodiment of his honour and intimate relationship. This clever transformation saved the farmer's status while displaying his hospitality, without damaging his commitment to his chicken or to his *shauq*. For rural Pakistani cockfighters, stories like this reiterate the import-ance of preserving the breed and bloodline of the fowls and highlight the value of the more-than-human relationship.

How can a deeper conceptualisation of "taste" allow us to comprehend the varied dynamics of more-than-human sociality? Put another way, how can an examination beyond "sight/vision" lead us to take non-humans as active subjects rather than as passive objects? To investigate such questions, it is useful to closely examine a set of emotions and attitudes (in the above case, these could be anger, frustration, and dissatisfaction) associated with taste. As we see, the poor farmer articulated his discomfort through "tasty" cooked chickens and conveyed his displeasure to the symbolic and somewhat direct violence of the rich over the poor. Hence, it may be argued that "vision" is not always the singular sense to shape people's everyday experiences (Stoller 1997, 3) and that "sensuous nonvisual elements ... constitute important aspects of the epistemology" of non-Western peoples (1997, xvi).

The taste of a gamecock's feed is also important, and its careful prepar-ation indicates how we can examine the strengthening of the human–cockerel bond in South Punjab through a sensuous analysis that goes beyond vision. When the mating season begins in spring, usually in March, the hen lays eggs and later remains with the chicks for four to five months, cockfighters like Baksha provide millet to the chicks twice a day. With more than thirty years of experience in breeding champion gamecocks, Baksha now stands at a respectful position in the cockfighting community of the area. His birth name, Allah Bux (meaning the gift of Allah), and his full-length silver beard can easily deceive a stranger into assuming he is a religiously inclined person, someone who spends his days and nights praying in the mosque and pre-paring for afterlife. Yet only within the cockfighting community, where he was famous as Baksha, everybody knows him as a true *shuqeen* of cockfight and an expert craftsman of the gamecock's feed. Baksha told me that when the chick turns into a cockerel at about six months of age, he is then fed a special diet called *khurāk* to build a strong and disciplined physique. To prepare this, Baksha peels forty water-dipped almond kernels and grinds them with fifteen pistachio kernels in mortar and pestle by adding some crystallised rock sugar, ten grams of butter, and one gram of black pepper. After leaving the mixture for five minutes, he adds ingredients that aid digestion, such as two green cardamoms and a few fennel seeds. At the last stage, he adds a small amount of candied apple jam and rolls the mixture into small balls or *laḍo*.

Preparing the special diet for gamecocks is not only an investment of time and energy but also involves considerable financial investment. The ingredients are expensive and may amount to 500 Rupees per day ($3). When discussing such high costs of chicken feed, Baksha and others resort to the saying: "*shauq da ko'ī mul nhīn*" (*shauq* is priceless), denoting an affective dimension that cannot be monetised (see Narayan and Kavesh 2019). They do not see their money as wasted, but as an investment where returns are not

Figure 3.3 A cockfighter showing *khurāk laḍo*, one is given to a rooster each day.

measured by rational calculation but through achieving broader social objectives like fulfilling *shauq* and demonstrating masculinity through fighting the cocks in the arena to accumulate honour and prestige. However, Bakhsha and other cockfighters hold that a pedigree rooster cannot be readied to fight on an expensive diet alone but has to be put through a strong regimen of training to develop stamina and to sharpen his reflexes. The embodied relationship between a gamecock and a man, developed through taste, enters into the next stage where the cock's body is touched and massaged for training purposes. This transformation from "taste" to "touch" is important as it helps the cockfighter sculpt a good breed into a champion rooster to ensure that he fights with courage and displays formidable qualities.

Pir Syed Qutab Shah, one of the most respected cockfighters in my field site, detailed the intricacies involved in training a rooster. I first met him in December 2014 at a cockfighting competition near Bahawalpur City. His short white beard, maroon embroidered Sindhi cap, and an agate ring worn in the right hand depicted what was obvious to many South Punjabi people, a sign of spirituality and noble descent. Referred to as Pir Sayen, he was in his early 50s and had nurtured his enthusiasm for fighting cocks for the past twenty years. Pir Sayen pointed to different fighting styles in cocks: some fought with their beak, others with spurs; some attacked instantly while others waited for a good moment; some attacked by flying high and others fought entirely from the ground; some were slow but strong while others were fast but could not bear a single blow. According to him, the mastery of cockfighters lied in recognising these different capacities in a cockerel and training him to boost his particular capabilities. Such skill was acquired through a deep interaction consisted on years of inter-species sociality, and knowledge of the cock's body. So, if a cock had pointy long spurs, a sharp beak and strong feathers, cockfighters usually trained him for spurs fighting to strengthen the ability for quick and sharp attacks, or if a cock had loosely grown spurs and preferred to fight from the ground, he was trained for covered-heel fight by improving his stamina, endurance, and power.

The exercise regimen delineated by Pir Sayen began with a fifteen-to twenty-minute morning run where he went jogging with the cockerel. Then, there was a jumping exercise (*bharī*) which, according to him, was a type of push-up for roosters, "as wrestlers do push-ups to increase stamina, likewise we give *bharī* to the cockerel to keep it in best physical shape." This exercise required almost half an hour and was believed to increase the stamina and power of the cockerel so that he was able to make flying attacks during cockfights. The training session was complemented by a dry massage at the end of the workout. Pir Sayen recommended that the cockerel should be dry-massaged on its wings, thighs, and neck for thirty minutes, as it "prevents fat accumulating over the rooster's body, and makes him strong like iron." After the massage, some trainers preferred to spray cold water through their mouth on to the cock's face in the belief that this will prevent swelling during combat. Some also fomented the head, the wings, and the thighs of the rooster with a salty wet pad to make the bird's flesh hard. In the afternoon, the gamecock was massaged again when cockfighters knead (*ghutna*) the bird's legs for half an hour.

Figure 3.4 Through massaging and training, the men develop a strong bond with their gamecocks.

This process of extensive training and massaging means, as I suggested earlier, that the cockfighter spends a good part of his day with roosters. Given that many enthusiasts are not people of means and need to work for a livelihood, some men take their roosters with them to work. For instance, a police constable may take his rooster to the police station, a vegetable vendor can keep his rooster under his cart, a school teacher can take a bird to school, and a bus conductor can hold it under his arms while haggling with passengers over daily fares. These are not rare occurrences. Of course, there are cockfighters who, for some reason, cannot take their roosters to work and try to return home as early as possible to spend the remainder of the day with them. Either way, the bird remains a crucial part of their everyday life and strongly guides their activities, priorities, and choices.

Once the cockerel is one year old, his training is intensified with sparring exercises (*tapānah, kashnā*). These exercises are a crucial *rite-de-passage*, where the bird is tested for about ten minutes against a senior rooster to give him a taste of the fight. Cockfighters closely observe whether their cockerel has increased in strength and stamina and alter his diet and training regimen accordingly. As the day of the cockfight gets closer, the training stops and the bird is given only regular massage and rest.

A careful understanding of the non-human body, developed through a sensuous relationship of sound, taste, and touch, enables cockfighters in assessing the cock's fighting spirit. The relationship generated through touch particularly helps the cockfighter develop an embodied understanding of the cock's body—his excitement for training and his fatigue after each training

Figure 3.5 Two enthusiasts dry massaging their rooster.

session. By understanding the rhythm of the cock's heartbeat, breathing, steps, and blinking, cockfighters can describe how the bird is responding to every touch and each stroke. Thus, a sensuous understanding of the cock's body helps a man develop and strengthen a meaningful relationship with the rooster, and take him as a subject who will showcase their hard work and expertise in the fighting arena. It is in the arena that roosters epitomise the masculinity of their keepers—masquerade in their power, physical strength, and fighting abilities—and enable them to achieve crucial symbolic rewards by outshining a strong opponent.

An analysis of cockfighting focused on the sense of vision is also important and can inform how the men contest masculinity through their roosters in front of their peers. As two roosters engage with each other in fierce combat, the more-than-human intimacy developed through breeding, training, massaging, and feeding is tested. Before the gamecocks are released in the fighting arena, men of all ages find a place to sit either on chairs or on the ground and surround the cockfighting pit from every angle. Only handlers are allowed to touch the fighting cocks, and everyone becomes quiet and attentive to see and judge the courage, bravery, and strength of the roosters. Thus, the cockfighting arena becomes a stage where the emotional bonding between a man and his rooster is on public display and where, through their bird's performance, each man is judged for his commitment by his peers. Below, I examine such cultural meanings associated with cockfighting in rural Pakistan.

Feathered clashes

"There are two types of cockfighting," explained Ghulam Hussain: "*khulā-kanḍā* [bare-heel fighting or spurs fighting] is like a sword battle and does not last for more than twenty to twenty-five minutes, whereas," he argued about *banḍ-kanḍā* [covered-heel fighting or beak-to-beak fighting] "the other is more like boxing where both roosters fight for almost forty to fifty minutes."

Ghulam Hussain, locally known as Ghamo, was a broad-shouldered man in his mid-40s with a thick moustache who lived in Munshi Wala village near Lodhran City where he had a small mango farm. Driving his old Suzuki Mehran, he took me to a bare-heel cockfighting competition at Teerah Soling village, near the city of Bahawalpur. I was excited and somewhat frightened to see the "sword fight" where roosters were expected to fight with their sharp spurs to defeat or kill the opponent bird and thereby humiliate his keeper.

Both type of cockfighting, "sword fight" (or bare-heel) and "boxing" (or covered-heel), were organised on Sundays in rural areas. At the time of my fieldwork in late 2014 and early 2015, there were frequent police raids on cockfighting pits after the appointment of a new District Police Officer who, according to Ghamo and other cockfighters, was a devout Muslim and considered cockfighting and gambling major sins. Remote rural areas were relatively safer for such fights because the police did not raid villages, and on the rare occasion when they did, the men could flee into the fields easily.

Figure 3.6 Layout of a South Punjabi cockfight.

Often the courtyard of a hospital or a school, both government buildings, was chosen to hold the event to confuse the police about the organiser. Holding a competition at a private residence was risky, since the revised version of the Gambling Act of Pakistan had enabled the police to launch an investigation against the owner. On Sundays, since rural hospitals (usually Basic Health Units) and schools were closed, cockfighters enjoyed the fight without worrying too much about external interference.

It took us almost an hour and a half to reach the village school where the spurs fight was being organised. *Shauqeens*, carrying their birds under their arms, were swiftly crowding the arena and greeting each other warmly. Many laughed while discussing the detailed preparation of their roosters and their plans for lodging the bets. They also spoke about the increased police raids. After saying Salam to some of Ghamo's close friends, I followed him as he found seats around the rectangular cockfighting arena. The audience soon surrounded the pit from every side and the "sword battle" began.

Khulā-Kanḍā (bare-heel) cockfighting

Razor-sharp metal spurs, pointed swords, blades, or spikes are not used in South Punjabi cockfighting. Rather, natural spurs (*kanḍā*) are considered the birds' "swords." They can be lethal too and are filed to needle-like sharpness. Watching the first bout, Ghamo noted that if the rooster was too young to have fully developed spurs (since it took a two-year-old cock to develop full-length spurs), external spurs from a dead cock were fitted to his heels with tape. One of the participating cocks had external spurs fitted to his ankles. The process of fitting the spur is done with extreme care by experts and once the combat begins, handlers are not permitted to retie a fallen spur. There is no rule over the length of the spur (usually spurs from 2½ to 3½ inches long are chosen); however, extra-long spurs are generally avoided because they can injure the cock.

South Punjabi cockfights are never declared a draw, and roosters keep fighting until a clear winner is declared.[3] However, there are different rounds (*pānī*), each lasting for six minutes, that provide an opportunity to participating cockfighters and handlers (*mahāwat*) to facilitate the fighting roosters. There are also two experts in the arena, one from each side, who provide a specialised mixture (*gālah*) to the gamebirds to enhance their stamina. Both experts and handlers are given some money by the competing cockfighters at the end of the fight as a token of their services. The average fee is around 300 Rupees ($2) for the expert, and handlers receive 500 Rupees ($3) per fight. In other words, the outlay for the activity is relatively low, at around 1600 Rupees ($10).

As Ghamo and I watched the first bout of this "sword battle," he extolled the beauty of the fight: how the cocks dodged attacks, looked for a good move, hit with the beaks, jumped high in the air, and hit with the spurs. What might seem an act of cruel chicken fight to an outsider, appeared quite beautiful to cockfighters like Ghamo. He recognised each rooster as a distinct individual who was bred, fed, and trained for this very moment, and was now fighting like a brave man, with fury and courage. It seemed like everything was happening in slow motion to him, and as a true *shauqeen*, he did not miss a single kick,

Figure 3.7 The process of fitting external spurs.

dodge, or blow. Sometimes, he would mimic the moves of the fighting birds to show how intelligent they were in parrying an assertive attack and baffling the opponent. There, in such a chicken fight, he found a pure delight and genuine pleasure. As one of the cocks attacked with his spurs, a roar erupted among the men backing it and the excitement grew. "*Bahut ache,*" (very good) "*shābāsh,*" (well done) "*zabardast,*" (superb) and other admiring superlatives were shouted from the audience. The birds continued to lunge at each other and tried to plant their spurs on the adversary's head, mingling their rustling feathers and dodging the blows until the referee (*munṣif*)[4] announced a break and the panting warriors were taken away by their handlers.

During the break, the handlers worked frantically to motivate the gamecocks and keep them in good shape for the next round. They did so by deploying a range of techniques, from sucking their beak and face to spraying fresh water on their body. Vitamins were injected, and the bird was massaged to relax the muscles. One bird injured his eyes and beak during the combat, and the handler performed a minor surgical operation to get him on his feet for the coming round. In this process, the handler's clothes were drenched in the roosters' blood, making the bird look like a part of his body. Supplementing the handlers, the experts at this stage provided a special mixture (*gālah*) to the roosters to revitalise their energy, and the eager fighters were deposited once again at the centre of the arena for the fight to recommence.

Figure 3.8 A bare-heel cockfight in progress.

The next round was far shorter: once the half-blind rooster with a broken beak received a fatal blow, it was all over. Like others, Ghamo jumped with joy, throwing his shawl skywards, hands in my hands, mimicking the final blow. The winner proudly kissed his battered rooster while the loser quietly collected the carcass and left. After about ten minutes, the handlers fitted the spurs on the next fighting pair, and the second cockfight began. The organisers had scheduled eight cockfights for that day, in the hope that each of them would last for roughly twenty to twenty-five minutes.

The third cockfight was disrupted when a man came running and announced, "The police is coming!" (*polīs āndī pa'ī he*). Almost 200 men were caught off guard in the school courtyard and pandemonium followed. Half of them ran in all directions, some on foot, others towards their motorbikes, while the other half (including Ghamo and I) stood paralysed by indecision. At the sight of three black police vans entering the main gate of the school, we managed to gather our wits and ran towards the western exit, jumping over the short school wall, crossing wheat fields, and finding refuge in a mosque, hoping the police would not search "the house of Allah."

While fleeing from the scene, all I could think about were two things: the Punjab police and Clifford Geertz. Running was the only option: as a Pakistani, I knew that no matter how many documents I presented as proof of my status as an academic researcher, my presence at the cockfight meant I was implicated and a hefty bribe would only clear my name—a standard practice in South Punjab. I also ran because I felt compelled to do so alongside my hosts, friends, and respondents, like Ghamo and other cockfighting enthusiasts. After that incident, I realised that my action reinforced their trust in me and the men came around and began speaking to me more often about their life, sharing their stories, and inviting me regularly to cockfighting

Figure 3.9 Cockfighters fixing the broken beak of a gamecock during the break.

competitions. Geertz's police chase at a Balinese cockfight reflects a similar experience:

> Getting caught, or almost caught, in a vice raid is perhaps not a very generalizable recipe for achieving that mysterious necessity of anthropological field work, rapport, but for me it worked very well. It led to a sudden and unusually complete acceptance into a society extremely difficult for outsiders to penetrate. It gave me the kind of immediate, insider-view grasp of an aspect of "peasant mentality" that anthropologists not fortunate enough to flee headlong with their subjects from armed authorities normally do not get.
>
> (1973, 416)

Ghamo and I remained in the mosque for half an hour and then walked out warily. We joined other cockfighters across the road, just near the water channel where the mud water stream ran quietly. We could see police hovering in the arena, who remained there for two hours and apprehended a dozen or so cockfighters. Travelling back, Ghamo criticised the organiser, "it is a matter of great disrespect of participants that the police raided this event." He said that the organiser should have made the arrangements earlier, by either bribing the policemen beforehand, or using his social capital to hold a trouble-free event. "You see, *shauqeen* come from a long way off to participate in cockfights," he continued, "I think participating in *khulā-kaṇḍā* (sword battle) is not safe anymore." Ghamo was right. As we will see below, more people bet on "sword battles" than on covered-heel or "boxing" cockfight, hence the police tend to raid these more often. After the police incident, very few "sword battles" were organised that year. I attended about a dozen more "boxing" competitions (*baṇd-kaṇḍā*) during my fieldwork and tried to identify some crucial differences between these two types of cockfighting.

The "sword battle," as I discussed above, is more about appreciating the speed, agility, and attacking manoeuvres of gamecocks. The "boxing" cockfight, as we will see below, is focused on stamina, power, defence, and endurance. This is why lightweight roosters are selected for the "sword battle," since they can jump continuously and attack the opponent with their spurs while in "boxing" cockfights, heavyweight roosters are chosen for their ability to hit with kicks rather than deploying their spurs. "Covered-heel cockfights are a true test of a rooster's abilities," Mooda explained. He worked as a vegetable vendor in the main market of Kahror Pacca town. Mooda was a shortened version of his birth name, Mehmood, and among cockfighter, this middle-aged man was known as an expert breeder and trainer of quality roosters. He explained that the "sword battle" was more about luck (*qismat*) whereas the "boxing" fight was designed to test the breed and true potential of a cock. He rarely participated in "sword battles," because the latter was, according to him, a *shauq* of impatient people.

Baṇḍ-Kaṇḍā (covered-heel fighting or beak-to-beak fighting) Cockfighting

The "boxing" cockfighting competition is also illegally organised on a Sunday in the courtyard of a government building. It usually starts at ten in the morning with six pairs fighting on the same day and each combat lasts for about forty to fifty minutes. Although gamecocks are sometimes severely injured in combat, I did not witness a cock dying in a covered-heel cockfight. The reason is simple: like a boxer wears gloves in the bout, a gamecock in this type of combat has his natural spurs covered.

When Mooda readied his rooster to fight, I accompanied him to a "boxing" match. The competition was organised at the *ḍerah* (a spare place for men to sit, outside the household) of Anwar Khan Baloch, a landlord and relative of the Member National Assembly (MNA) of a district in South Punjab. Carpets and chairs were arranged at the centre of the *ḍerah*, where about 100 men sat expecting to watch a thrilling competition. Perhaps because it was the *ḍerah* of a politically influential person where the police was not likely to come unannounced or maybe because of my previous experience of successfully fleeing from the police that gave me a tremendous boost in confidence. At that time, participating in cockfighting with fearless enthusiasts, I felt ready to gain my interlocutor's trust, and most importantly, to witness something unattainable to thousands of rural villagers living just outside the four walls of that *ḍerah*.

Mooda's rooster, Jāṇbāz, was selected to fight against Javeed's rooster as the third pair. When the first two pairs finished fighting, Mooda and Javeed were asked to bring their fighters into the arena. Both roosters were dry-massaged and then given to the handlers for combat. It was half-past one, the boisterous

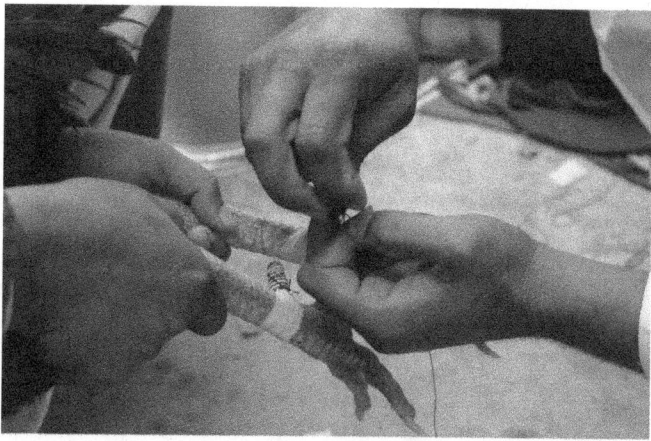

Figure 3.10 Men covering the natural spurs for a covered-heel cockfight.

audience of almost 100 men fell silent as the fight commenced. Both roosters started pecking at each other's face and head, and attacked aggressively by leaping into the air. When the birds finished the first round, Mooda instantly grabbed Jānbāz and started sucking the bloody head and neck of the animal with his mouth.[5] Later, he sprayed salted water on Jānbāz's face and body and placed a warm pad on the cock's face. The special dietary mixture was offered to the bird in the hope that it would save Mooda from the humiliation of loss.

About ten minutes later, Jānbāz was lowered into the arena to face his adversary for the next round. After two more rounds, Jānbāz gave up and fled from the fight. This was the most shameful way of losing, compared to the other, less shameful ways of losing a battle, such as when a bird surrendered by ceasing to fight, or he was picked up by the keeper because of a severe injury. However, if a rooster fled, the keeper and his supporters were disgraced (*beztī*) for it. The basic purpose of staging cockfighting, some *shauqeen* argued, was to appreciate the perseverance of cocks despite injuries. However, when a bird fled from the fight, it undermined the idea behind the cockfight. This act of fleeing displayed cowardice and "non-masculine" qualities of the gamecock, and raised doubts on the training of the keeper. Such views seem to have universal appeal, as Jim Harris (1994) while writing about some universal rules of cockfighting, stated that "a runaway chicken is hated by all cockfighters, and a runaway can never be a winner" (1994, 13).

Figure 3.11 The fighting cocks are facilitated by their specified handler during the combat.

As Jānbāz fled from the scene, we were all implicated in the loss, feeling disappointment and shame as his supporters. The opponent party ran into the arena, the men shouting in joy, hugging each other, clapping, cheering, and mocking Jānbāz, and us by extension. A person from the crowd approached us, blaming Jānbāz, "*aṣīl* cocks never run away from the fight, no matter if they are kicked in the face a hundred times." He and some other men from the opponent party argued that Jānbāz was not a true *aṣīl* but a bad breed (*badnaslah*). Another man recalled the final blow, mimicking how Jānbāz fled with "120 km/hour" speed to avoid death.

In South Punjab, the honour and prestige of an enthusiast lie with the performance of his gamecock rather than the end-result of the fight. A rooster that fights a hard battle and loses the combat but does not run away is acknowledged by the audience for his courage and bravery. Losing or winning a single fight does not significantly alter the keeper's status among peers, what counts is the manner in which the success or defeat occurs. Even though temporary humiliation always follows a defeat, past successes are valued. However, this loss was a bitter one. As I stayed around Mooda to support him in his humiliation and listened to the jibes, I understood what it felt like to be involved in a cockfight where one's *izzat* is at stake. It became increasingly clear that the cockfight is not merely a fight between two chickens or a competition to win money through wagering; it was a public display of masculinity, where honour and prestige could be gained or lost.

The cockfighting arena was a rare opportunity for enthusiasts to compete against men of different classes and win *izzat*. Despite the risk of losing honour, the cockfighting arena provided a sense of camaraderie to the men and a stage for receiving applause and recognition from peers. If Jānbāz had won, the script of the story would have been different. There would have been ten minutes of celebration, people would have showered praise on Jānbāz and Mooda, and I would have shared in the joys of status-elevation – albeit temporarily. This was not the case, and following a "shameful" defeat, two silent men and one injured rooster were heading back east, leaving the dusty arena and an orange sky behind.

As stated earlier, when a cockfight begins—either a "sword battle" or a "boxing"—the sense of vision takes precedence. Everyone observes the fight keenly to scrutinise any tiredness and exhaustion shown by the rooster and to praise any outburst of energy displayed in flying attacks. As the men focus their gaze on the fighting cocks' bodies, they start seeing them as warriors out to create something beautiful through the masculine display of combat. To my eye, the sight of a cockfight was also meaningful. Initially, the injury to the animal was a disturbing sight but gradually, as I forced myself to participate in more cockfights, it turned into a subject of study. I somehow learned to distance my emotions from the sufferings of fighting fowls and to observe them as a scientist looks at a laboratory mouse before performing a potentially lethal experiment on him. However, it was not simple. The birds were real, their blood was real, and the fight was real. I knew some of these birds and watching them get seriously injured was difficult. All through my fieldwork, as I played hide and seek with the reality of cockfighting, it became clear to me that I was never going to be a cockfighting *shauqeen*.

Before sharing my observations on the relationships between cockfighters, it is important to note that in this sensuous relationship between a man and a cock, the sense of smell is always active. The smell of a roosters' feed, their droppings in the cage, the smell of the fighting pit with the earth freshly showered with water, and the smell of human and animal bodies, of carrying a rooster under your arm, and of hugging a friend in wild joy after the cock-fight are meaningful to everyone involved in this *shauq*. Then there are some other smells, such as the smell of cowardice, or of courage and aggression; and the smell of defeat, of not being a "true" man. These smells, sometimes mingle with each other to create a distinct aura to help a cockfighter take his pursuit seriously and to develop a deep relationship with his birds, as well as with his friends who vigorously support him during cockfighting.

Friends, brothers, and gamblers

Malik Khizar, the man who told me that he would not give his pedigree chicken even if the Prime Minister of Pakistan were to ask him for it, had quite a different approach when it came to other cockfighters. He said, "If a poor, cockfighting brother requests this of me, I will oblige him without a second thought." As discussed earlier, Malik Khizar was an affluent man and a landlord of the village Khan-da-Kho, yet this never stopped him from befriending lower-class cockfighting enthusiasts. To him, the *shauq* of cock-fighting could not be undermined by external considerations of class, sect-arian, ethnic, or other social divisions. Rather, he stressed the egalitarian nature of this *shauq*, where a man's economic and social status was irrelevant. "I would prefer the company of a poor cockfighter than the MNA (Member National Assembly) of district Lodhran," he later stated. Such statements are common among cockfighting enthusiasts of the area and are reflective of a shared ethos that bonds them first and foremost as *shauqeen*. Indeed, that is the nature of *shauq*, as I argue throughout this book, that it transcends everyday societal boundaries (see Narayan and Kavesh 2019).

To signal such transcendence, South Punjabi cockfighters refer to each other as *bhirā* (brother). Within this brotherhood (*bhirapī*) of cockfighters, all social differences are meaningless and only the purity of *shauq* is considered crucial. Their *shauq* allows them to get together to form a community of shared interests and exchange ideas about the feeding or training techniques. In the arena, this brotherhood becomes more explicit. The enthusiasts, irre-spective of their social and economic differences, greet each other affec-tionately and sit together on the ground to watch the fight. Such practices coincide with the ethnographic observations of other South Asian scholars who have studied *shauq*-like activities. Kirin Narayan, for example, in her study of the women singers of Kangra, notes that because of their *sukinni* (or *shauq*) of singing, many "sisters" sit together in spite of their caste and other social differences (2016, 98–102). Clare Wilkinson also observed the egalitarian nature of such communities of common interests, and noted that both Hindu and Muslim women embroiderers of Lucknow engage in the *shauq* of embroidery, and religious differences are not pronounced (1999, 133). Such observations reaffirm that the sharing of a common *shauq* allows people to form intimacies that go beyond established social categories, even in societies structured on class, caste, and religious disparities.

However, if the *shauq* of keeping and fighting cocks is seen as pure enthusiasm requiring long-term commitment and genuine passion, then the pursuit of money through this activity is seen as one bearing tensions. If a person participates in a cockfight only for the money, he is believed to lack "genuine *shauq*." For instance, Imran Niazi accused another cockfighter, Baksha, of being money-oriented, saying that "he only participates in cockfight for the love of money (through betting), and does not have real *shauq* of cockfighting." Similarly, Waso scoffed at Malik Khizar, arguing that "he keeps chickens to sell their chicks, and does not have the courage to fight his roosters." Profiting from the *shauq* is frowned upon. I will have more to say about this in Chapter 6, where I examine in detail the criticism and tensions that arise when money and *shauq* are mixed, but here this complex interrelationship begs the question: if profiting from *shauq* is criticised then how can the explicitly monetary and morally derided activity of betting (*jowā*) be so central to the practice of cockfighting?

The answer lies in the "rules" of cockfighting. Like in most places in the world, a cockfight in South Punjab proceeds with betting. According to my interlocutors, betting serves two purposes. First, it increases commitment to the fight. When a man bets on fighting cocks, he bets a part of his honour and thus takes the bout as a serious pursuit rather than as mere fun. Second, betting allows friends to display their loyalty and allegiance to the cockfighter. This turns betting into a carefully considered practice, underpinned by cultural ideas about the seriousness of play and the continuity of social relations.

The cockfight commences with a wager, and both cockfighters collect money from their friends and submit it to the referee. According to my calculations, from a sample of almost eighty cockfights, the average bet on a single pair in a covered-heel (boxing) cockfight was 10,000 Rupees (about $60) and in bare-heel (sword battle) cockfights, it was 25,000 Rupees (about $150). The difference in the betting amount in both types of cockfighting was marked by the duration of the combat. In bare-heel, the rooster fought for less time and attracted more wagers, while in covered-heel, the contest went on for many rounds and inspired the attention of the audience towards the fight rather than the result (or the money).

Betting is done collectively, combining small amounts from many friends. A man cannot simply appear in the arena with his gamecock and money and demand to participate in a fight. To participate, he needs the support of his friends who can back him up and wager on the fighting rooster. In other words, money cannot buy an entry ticket into the arena, nor can a prized cock; it is a necessary but not sufficient condition of participation. The bets are there as a public demonstration of a man's social relations and endorsement of fellow *shauqeen*.

What is equally intriguing is the cap on the amount generally considered for such bets. Waso informed me that people are discouraged to place more than 1000 Rupees (about $6) on a cock, "Cockfighting is not about making money; money is there only to boost the thrill and joy of the fight." This may also be related to the egalitarian nature of this *shauq* where men are praised for not displaying wealth and power, or rendering the event overly commercial. In other words, the rules around wagering ensure that the potentially corrupting influence of money is curtailed and does not impinge on genuine passion. Indeed, this regulated form of betting reinforces the notion of *shauq* by showing restraint and consideration, rendering money as more of a prop than an end in itself. Hence, people constantly reaffirm that a man should never be motivated by profit while engaging in *shauq*. To my surprise, there were no dishonoured bets. This happened because all the audience comprised cockfighters and everyone honoured the rules of engagement. This was in line with the local norms and values: an honourable man would honour his word. If a man refuses to pay his bets, he not only risks losing his participation in the cockfighting group, but faces social exclusion and derision for being unreliable, dishonest, and ultimately not a "real" man. In other words, cockfighting and betting reinforce values that underpin the foundation of masculinity in South Punjabi society.

A contest for honour and masculinity

The practice of fighting roosters and its association with masculinity can be analysed, as folklorist Alan Dundes (1994) does, by drawing on psychological approaches. In this form of analysis, the cockfight becomes "a thinly disguised symbolic homoerotic masturbatory phallic duel, with the winner emasculating the loser through castration or feminization" (Dundes 1994, 251). However, such an approach seems to be focused on individual traits, favouring the unconscious, with little consideration of the ethnographic context, symbolic gains, and social relations. My account of South Punjabi cockfights avoids the Freudian approach and makes no attempt to uncover the unconscious motivations underlying the activity. Rather, I have focused on the cultural interpretation of cockfighting through a sensuous analysis and explored how culturally meaningful ideas about what constitutes masculinity and honour are exchanged and displayed in the arena. This interpretive approach, to follow Geertz (1973), relies on thick description of both processes and events to foreground a way of life and social practices. What I found important, and what was emphasised by my respondents, was the role of this *shauq* in the struggle between men to achieve honour based on a display of hyper-masculinity. Thus, my analysis explores how cockfighting in South

Punjab is a gendered activity, largely confined to the male domain, where masculine honour is reproduced and modified by fighting cherished birds.

Such behaviour is visible in other communities of enthusiasts. For instance, Wacquant (2004; 1995) while writing about boxing in Chicago observes that, in the ring, boxers achieve what they may not be able to achieve outside of the boxing arena—money, fame, respect. The enthusiasm for boxing, which Wacquant explains as "organic connection ... akin to a religious allegiance," and "a form of possession" (1995, 507), elevates the status of the boxers among their peers and take their practice to define "their innermost identity, their practical attachments, and everyday doings, and their access to and place in the public realm" (1995, 507). Similarly, cockfighting, allows the South Punjabi enthusiast to achieve important symbolic rewards such as *izzat* through the masculine performance of their birds, who fight like boxers in the pit.

Men in South Punjab are expected to act in an assertive and demonstrative manner, cautious not to display behaviours associated with women, such as shyness or avoiding public gatherings. The "true" male (*juoān mard*) is expected to have three main characteristics: courage (*jur'at*), bravery (*dilerī*), and strength (*zor*). These attributes are considered crucial for a man to surmount life's many challenges, such as being able to provide for and protect his family, display courage in times of hardship, and defend his honour from all threats. Similarly, a cock is expected to stay true to his nature and not flee the fighting ground during the cockfight. If a gamecock opts to run away from the fight, he is believed to have behaved like a hen. In the fighting arena, the cock thus becomes the surrogate of his keeper and is expected to embody and display male qualities. So, when a cock fights with courage (*jur'at*) and without fear (*dilerī*), and remains strong (*tāqat-war*) despite injuries, his keeper is acknowledged for breeding, feeding, and training a truly masculine rooster. However, if the cock fails to attack, fights without passion, or flees from the arena, this reflects badly on the keeper, who is then subjected to humiliation by the audience. To an extent then, the cockfight arena is a site for the reproduction and assertion of masculine traits, where the symbolic capital associated with *izzat* can be gained or lost.

In the arena, the cockfighter is also expected to display characteristics of a "true" male. This means that he must relinquish the emotional attachment to his rooster and stay calm in critical times, encouraging the rooster to attack despite serious injuries. If a man shows hesitation in subjecting his cock to a fight, withdraws from the contest prematurely, or complains about the advantages of his opponent, he is not considered manly and such acts are remembered and recalled repeatedly among his peers. In the arena, a cockfighter is expected to wear the mask of hegemonic masculinity, and without showing any sign of weakness, exhibit the dominant qualities of South Punjabi masculinity through his behavioural and bodily display.

In sum, after taking its lead from Geertz's seminal study on the Balinese cockfight (1973), this chapter utilises a sensuous analysis to expand on

Figure 3.12 A runaway cock is mocked and humiliated along with his keeper.

different modalities of the human–cock relationship in South Punjab. The sensuous ethnographic approach guides us to explore the sound, taste, touch, smell, and sight of gamecocks to uncover the human–cock relationship through breeding, feeding, and training roosters, and document the pains and pleasures involved in the *shauq* of cockfighting. Such an approach also enables us to examine how cockfighting enthusiasts tend to see gamecocks as endowed with distinct personalities, combat styles, and histories of success or failure, and why the men display dominant traits of South Punjabi masculinity by fighting their cherished birds in the arena to gain *izzat*—the main ingredient of a man's symbolic capital in rural South Punjab.

In the masculine arena of cockfighting, the audience, supporters, and friends congregate to praise the combative spirit of the winner gamecock and humiliate the cowardly chicken that flees from the fight. They bet on the rooster and expect him to fight with the courage (*jur'at*), bravery (*dilerī*), and strength (*zor*)—three basic qualities associated with the "true" male in South Punjab. When successful, the cock not only increases the status of his keeper but also helps him build solid friendships that transcend class and other social differences. The *shauq* of cockfighting then becomes a performance where hegemonic masculine traits are reproduced and rehearsed, and where prestige is gained and lost before peers. The activity also showcases different values that underpin inter-personal relationships in rural Pakistan and, along with dogfighting which I discuss next, allows us to construct a picture of rural

South Punjabi culture where the accumulation of *izzat* is considered the ultimate objective of a man's life.

Notes

1 The word *kukaṛ* or *murgh* means a "cock," while the suffix *bāzī* in cockfighting is strictly used for gambling.
2 The five prayers are *Fajr* (dawn), *Dhuhr* (noon), '*Aṣr* (afternoon), *Maghrib* (sunset), and *Isha'a* (night). Among all these, *Fajr* is the most trying prayer, as it requires waking from sleep. The rooster's cockcrow at dawn helps in waking the faithful. Therefore, the textual traditions in Islam and folk traditions of South Punjab admire the bird for this quality.
3 In other parts of the world, see for instance Marvin (1984, 64), cockfights are sometime declared a draw after thirty minutes of fight.
4 *Munṣif* (literally, a referee) is a timekeeper in cockfighting. They are respected cockfighters whose authority or decisions are rarely questioned.
5 Smith and Daniel (1975, 86) have called this "beak-to-mouth resuscitation." It is believed that this practice is found everywhere among cockfighters.

4   The spectacle of dogfighting

Amplified masculinity

In October 2014, I met Makhan Pehlwan, who told me about a dogfighting competition that was going to be organised in Hussainabad. A muscular man in his early 30s, Makhan had an interesting story attached to his name. His birth name was Iqbal and, being a well-built and healthy boy, his father Ashiq decided to make him a wrestler (*pehlwān*) and named him Makhan (literally butter). As local wrestlers consume large quantities of butter (*makhan*), *ghī* (a form of clarified butter), almonds, milk, and other protein rich diets; his name reflected his soon-to-be status. However, despite his strong diet and laborious training regimen, burly Makhan Pehlwan never won a competition in his wrestling career. Considering his persistent cold streak, his father ceased giving him butter and diverted effort and resources toward his younger son, Abid. Unlike Makhan, his brother Abid became a successful wrestler and won many bouts. Apart from wrestling, the men of the family had a longstanding interest in dogfighting and busied themselves with training the family dog for upcoming dogfights.

Drawing a parallel between dogfighting and men's wrestling is common in rural South Punjab, particularly among Seraiki-speaking people. First, because both fighting dogs and men are fed and trained in similar fashion and, second, because in both dogfighting and wrestling several of the traits associated with hegemonic masculinity are reproduced. Before a large audience, both dogfighters and wrestlers try to gain symbolic resources such as honour and social prestige by displaying their strength and courage to engage in combat. Makhan might have been unimpressive as a potential wrestler, but his extensive knowledge concerning the sculpting of an athletic body made him a popular trainer of fighting dogs. When he came to know that I was writing a book on dogfighting, he became my guide and encouraged me to participate in a famous dogfighting competition in Hussainabad.

As the date of the competition drew nearer, I found an aura of enthusiasm enveloping dogfighting aficionados. The event was announced on local radio and news of the fight circulated through rural networks. On the given day, I accompanied Makhan on his motorbike and we travelled almost thirty kilometres towards east, along cotton fields and then across the River Sutlej in a boat until we reached the local festival grounds (*mela*). Dogfighting was organised as a part of this festival on a wide arena, encircled by about 5000 men of all ages and social classes. Elbowing our way forward, we soon reached the front row and there, for the first time in my life, I witnessed a dogfight— two tall, blood-smeared dogs locked in fierce combat.

Makhan's family dog was not taking part in the competition but he took his mobile phone out and set the timer, telling me how excited he was to watch and scout (*ḍaikhan sambhālan*) other dogs by noting the duration of the fight, and evaluating their stamina and endurance. To my surprise, Makhan was familiar with the winning history of both fighting dogs, and even indicated a dog who, according to him, had raised the hairs on his back and was about to lose the contest. He was right. In another couple of minutes, the dog Makhan had identified as the loser, yelped and gave up the fight.

All of a sudden, the silent crowd became boisterous. The winning dog-fighter raised his dog to his shoulders and drummers (*ḍhol* players) along with other local minstrels, played the joyous melodies of celebration. While loudspeakers blared praise for the winner dogfighter, "dog of Sajjad Joya has won the combat (*Janāb Sajjād Joy'ā dā kuttāh muqābalah jit gyā*)," a friend provided a red 100 Rupee bill to a rose garland seller and placed the garland around the dogfighter's neck, something like a medal of honour. Other friends gave money (*veyl*) to the announcer, drummers, and minstrels as a reward and then joined the celebration by dancing the South Punjabi *jhummar*, raising their arms and synchronising legs. Holding the leash of the winner canine who was now exhausted and panting, the dogfighter and his ten-year-old son proudly walked around the arena and received applause from the audience. This celebratory performance was directed towards the defeated dogfighter and his friends who were quietly leaving the arena. Within ten minutes, all celebrations came to an abrupt end as the next combat began.

What are the characteristics of the entangled relationship developed between a dogfighter, the fighting dog, and the audience? What symbolic values are associated with the activity of dogfighting and how do they affect the social life of dogfighters beyond the fighting arena? My first question guides me to navigate the meaning of dogfighting and its significance to both humans and the non-human animal. The inter-species relationship developed between a dogfighter and his canine is based on years of intimacy, close attachment, and careful labour. Despite strong religiously perceived restrictions on keeping a dog in rural South Punjab, the dogfighter feeds and massages his dog, so he may appear strong and fearless to the large audience. As the dog enters the arena, it becomes a stage where the audience observes and judges how the dog reciprocates to the attention and care provided by his keeper.

The celebrations that follow a dogfight are an interesting lens to consider my second question about the social impact of a dogfight on the keeper's life. As I suggested in the Introduction of this book, dogfighting to some animal welfare groups in big cities like Lahore, Karachi, and Islamabad appears as an unethical, unjust, ultra-brutal practice, something close to insanity. It has been suggested as a trait "typical" of rural Pakistan, and particularly of Seraiki-speaking people of South Punjab, which reflects their lack of education and employment opportunities. Similarly, to a powerful Muslim cleric of the central mosque or madrassa, dogfighting is a sinful practice that involves an "impure" and "unholy" animal, and shows the illiteracy of the older generation. Yet, the organisation of such fights as a part of the festivals and its associated celebrations show that dogfighting is more than a sign of rural South Punjabi backwardness and may be considered a cultural practice entangled in religious and social discourses.

As this chapter shows, dogfighting is a *shauq* of diverse social classes, ranging from wealthy politicians, landlords, and village chiefs to salaried police officers, school teachers, and veterinary doctors, to those who own small businesses like tailors, small farmers, and livestock keepers. Like pigeon flying and cockfighting, dogfighting is an all-male activity; organised by men, watched by men, and participated in by male canines. Yet, it is different from pigeon flying and cockfighting in two key ways; first, it does not involve gambling (or prizes), second, the activity is organised at village festivals (*melas*). Such exceptions turn dogfighting into a display of honour and social prestige through a public performance of masculinity. I suggest that the dogfighting arena is a site for enacting South Punjabi masculinities where traits like courage (*jur'at*), bravery (*dilerī*), and strength (*zor*) are celebrated whereas retreat, surrender, and hesitation are mocked as cowardice. Drawing on the two questions posed above, I argue that the human–canine relationship in South Punjab becomes explicit as a type of *shauq* that expresses and validates the hegemonic masculine traits (*mardāngī*) and helps a man accumulate honour and prestige (*izzat*) by competing against village rivals.

Dogfighting is not only a combat between canines to establish supremacy but is also the struggle between men to publically compete for prestige. It is a conspicuous performance (Schechner 1993, 20) aimed at entertaining the audience, and to some extent displaying a man's symbolic power in rural areas. By participating in dogfighting, people admire the courage of the canines, their bravery in the fight, and the strength that allows them to keep fighting despite injuries. They expect the dog to be fearless like a "true" man and represent his keeper's *izzat* (honour) in the confrontation. In fact, dogs are expected to fight like human wrestlers in using their skills and technique to disempower the opponent in physical combat. By drawing on a comparison with wrestling, I take dogfighting as a productive lens to examine how crucial South Punjabi masculine attributes are rehearsed, reproduced, and reaffirmed.

The chapter further explicates how the passion for dog keeping influences the social lives of dogfighters and reinforces a value system marked by care, hard work, discipline, and domestication. One of the main differences between dogfighting as *shauq* and the other two discussed above is that it is centred around a *haram* (polluted) animal. By examining the canine's material and cultural status in South Punjabi Muslim society, I argue that the practice of keeping and fighting dogs allows the men to compete against their peers for *izzat* and status, with the aim of demonstrating their mastery, skills, and

Figure 4.1 Dogfight celebrations.

Figure 4.2 Dogfight celebrations.

expertise to a large crowd in the public sphere through an animal that, despite being ritually polluted, is expected to reflect manly qualities.

Dog—a *Haram* animal

In urban Pakistan, dogs are subject to contradictory forms of care. In parts of major cities like Karachi, Lahore, and Islamabad, keeping a dog signals participation in emerging upper-middle-class sensibilities where the animal is loved, petted, and even cuddled. Yet, away from the elite suburbs, dogs are kicked, beaten with sticks, and are generally victims of abuse. The most common reason for this attitude towards dogs is their perceived tension with local Islamic interpretations.

While pigeons and roosters are *halal*, and keeping them in the house and feeding them is considered a virtue, the dog is a *haram* animal and is considered ritually impure. *Halal* in Arabic means lawful and permitted, and Muslims are allowed to eat *halal* animals. *Haram*, however, is used for what is forbidden or prohibited, and *haram* animals are considered ritually inedible. The distinction between halal and haram cannot be understood through the taxonomic schema that Mary Douglas (1966) uses for kosher and non-kosher animals, explaining how, in Leviticus, cloven-hoofed animals that chew the cud (such as cattle, goat, and sheep) and fishes with fins and scales are kosher because they come under the symbolic order of what is pure, proper, and holy, whereas animals that are neither cloven-hoofed nor chew the cud (such as the pig, the camel, the hare, and the badger), and fishes without fins and scales are regarded as impure, improper, and thus unholy as they do not fall under the cherished order. The symbolic order of *halal* and *haram* animals in Islam is discussed in the major Islamic texts—the Qur'ān and Hadith—that forbid the meat of dead animals, predators, pigs, cats, dogs, and donkeys. There are diverse interpretations of the status of other animals in *Shia* and *Sunni* schools of thought, where, for example, some people consider the hare as *halal* while others avoid eating the animal.

In the rural areas of South Punjab, the interpretation of *halal* and *haram* is mostly culturally influenced where some *haram* animals like pigs and dogs are considered an embodiment of ultra-pollution, while others like horses, cats, or parrots are not. For example, in South Punjab, even uttering the word *sowar* (pig) is believed to void forty days of worship. Although there is no supporting evidence for this in Islamic texts, such cultural interpretations lead us to consider that those *haram* animals that do not fit into the symbolic order of rural South Punjabi cultural system are considered "matter out of place," and thus become extra-polluted (Douglas 1966, 36). The focus of this chapter, dogs, are considered the absolute embodiment of pollution and dirt. The thought of developing an intimate relationship with a dog, for instance, petting or cuddling him, feeding him dairy products (or "pure" things such as milk, butter, *ghī*), and spending time with him, goes against South Punjabi values. It is believed that if a dog sniffs or licks a cooking pot or vessel, it becomes unclean and needs to be washed with water seven times before being used again. However, after a lengthy discussion with Islamic scholars based at a madrassa in the town of Kahror Pacca, I learned that it is permissible to keep a dog for hunting or guarding purposes. A functional explanation for this is that, in traditional Arab society, dogs performed these two crucial functions and it was thus permissible for them to be domesticated for these instrumental reasons. Meanwhile, the dogs' saliva (and bite) was believed to carry disease (such as rabies), and again for health-related reasons dogs were discouraged from being kept inside the house. Two modes of understanding are aligned here: the religiously constituted conceptualisation of purity and pollution, and systematic reasoning of hygiene and health. When combined, these modes of understanding turn the dog into "matter out of place" and, as a polluted creature, place it at the margins of the South Punjabi cultural system.

Historian Alan Mikhail suggests a different line of argument and contends that dogs' status in Muslim societies altered with the emergence of modern cities. Citing Al-Ajhuri, an Egyptian Muslim scholar of the seventeenth century, Mikhail provides examples of Prophet Muhammad praying in the presence of dogs and how many of his companions owned and raised puppies (2015, 79–80). He suggests that besides their use for herding, hunting, and protection, dogs consumed garbage and kept the streets clean in many Muslim dominant cities (2015, 82). For this reason, they were provided with food and protection and those found guilty of violence against them were punished. However, in the early nineteenth century, as garbage started to be considered a threat to public hygiene in quickly urbanising societies, the dog began to be seen as a noise polluter, a source of filth, a potential vector for diseases, and a menace to the social order (2015, 85–88). Mikhail argues that this relatively recent change in Muslim's attitude towards dogs is now dominant in most Muslim societies and people believe that the dog was always considered polluted in Islam (Mikhail 2017).

Mikhail's argument about the positive historical status of dogs in Islam is supported by the Sufis who appreciated the animal's loyalty and fidelity in their poetry. Many of them argued that in the Qur'ān, dogs have not been described as a dirty or polluted animal rather, their positive attributes, including loyalty and companionship, have been extolled for instance in chapter 18 (The People of the Cave). Therefore, those who support the Sufis' point of view hold that Allah appreciates dogs as loyal companions to humans and has given them an elevated status.

Javad Nurbakhsh, an Iranian Sufi writer, in his work *Dogs: From a Sufi Point of View* (1989) defends the canines' status in Muslim societies. He argues that the Sufis were the first to show people that "the dog possessed virtuous qualities, qualities which many human beings, regarding themselves as the noblest of God's creatures, lacked" (1989, 4). He narrates a story from the Prophet's accounts about a prostitute who felt compassion for a dog panting from thirst. The woman made a rope of her scarf, tied it to her shoe, and drew out some water from the well to quench the dog's thirst and for this act of kindness, she was promised a place in paradise (Nurbakhsh 1989, 28–29; see also Fakhar-i-Abbas 2009, 38). The symbolic parallel between the prostitute and the dog is important here, since the prostitute is also considered a symbol of impurity and an embodiment of moral, physical, and occupational pollution. Perhaps it is her marginal position that made her empathise and understand the revulsion and neglect with which the dog was treated. The Sufis narrate this fable to highlight the value of a selfless act and compassion to animals.

Punjabi Sufi poets have also appreciated the fidelity and selfless characteristics of dogs. Perhaps the most famous of them is the short poem by poet Bulleh Shah (1680–1757), where he appreciates dogs for their devotion, exceptional dedication, and their superhuman qualities:

> You stay up at night and pray
> Dogs remain awake at night too, yet higher than you
> They never cease barking
> Sleep on a pile of trash, yet higher than you
> They never leave their beloved's doorstep
> Even if beaten with shoes, thus higher than you
> Bulleh Shah! Grab some spiritual wealth
> Or dogs supersede you
> Higher than you

Despite such writings, in the Sunni dominant Muslim society of rural South Punjab, the canine remains a polluting animal that is best kept away from the house. In this contradictory situation where developing a close relationship with dogs goes against local values, it may be asked how dogfighters are able to keep, breed, and train dogs. More importantly, how do fighting dogs become a symbolic expression of their keeper's *izzat* and *mardāngī* (manliness) despite being considered ritually and culturally reprehensible?

The answer lies in everyday connection with individual dogs. For dogfighters, the impurity of dogs as a *haram* animal is a part of their belief and, like many other Muslims, they consider the dog's body as ritually polluting. However, while accepting the dog's *haram* status in Islam, they also believe that not all dogs are alike. They argue that the behaviour of stray dogs is different from domesticated ones because, through their training, domesticated dogs come to accept the abstract boundary between the human and non-human world. For instance, they say if you give a piece of bread to a stray dog, the next day he will come searching for food in human dwellings whereas a domesticated dog will never touch the food without permission. This lack of behavioural training in stray dogs means that dogfighters place them at the bottom of the symbolic hierarchy of canines. Among domesticated dogs, there is further categorisation with fighting dogs on top of this symbolic hierarchy because they represent their keeper with masculine courage, aggression and strength.

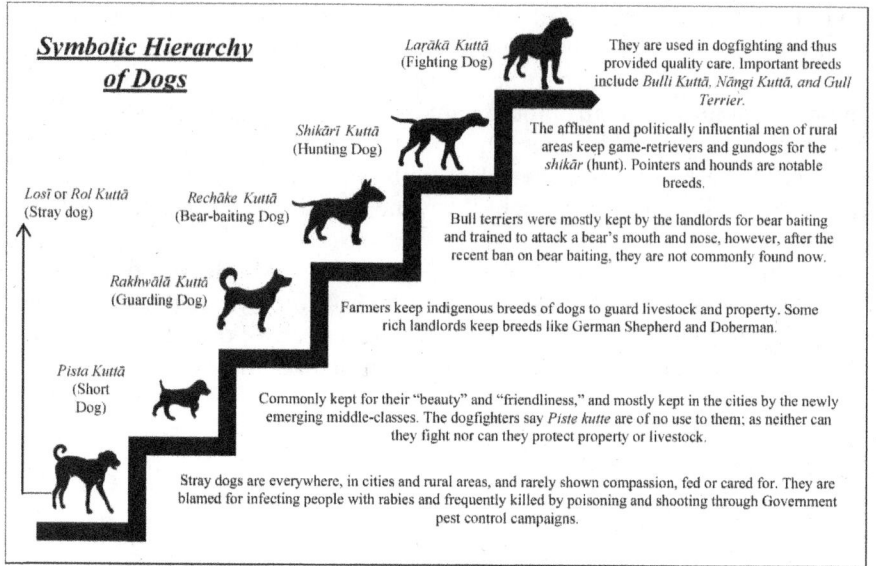

Figure 4.3 Hierarchy of dogs in South Punjab, as described by dogfighters.

In this hierarchy, fighting dogs are highly valued while stray dogs are despicable creatures. As stray dogs remain on the outskirts of villages and scavenge from waste dumps, their "undomesticated" nature and "wildness" make them a threat to human society. To tackle this threat, they are killed in large numbers on the orders of the Municipal Government by strychnine poisoning and shooting to combat the increasing cases of rabies and to maintain public safety and order.

The hunting dogs and bear-baiting dogs are mostly kept by affluent people with political connections who practise these illegal activities with impunity. For bear baiting particularly, people keep bull terriers and stage a fight against a tethered bear to display their material and symbolic power. In Chapter 1, I have discussed how the British introduced bear baiting and dogfighting into the subcontinent in an attempt to develop closer relations with local elites. However, after the ban on bear baiting by the Wildlife Department to protect the endangered Asiatic black bear, bull terriers are becoming hard to find in South Punjab. They are now being replaced with gull terriers that are specifically bred for dogfighting. In general, hunting dogs and bear-baiting dogs are higher up in the hierarchy of canines because they are kept by the wealthy and the powerful, while other domesticated (guard) dogs kept by villagers fall lower in this typology.

Guard dogs are used to protect property and livestock and are generally allowed inside the house. However, they are specifically trained to avoid contact with foods prepared and consumed by humans. To live in harmony with human cultural values, they are specifically trained to act and behave differently from stray dogs. The short dogs (*piste*) are the least favoured by dogfighters since they protect neither property nor livestock and cannot fight bears or other dogs. Such "unworthiness" makes them stand at the bottom of the symbolic canine hierarchy, just above stray dogs. However, given the rise of companion dog culture in some urban areas, many nuclear families now keep them as a status symbol and for what they consider to be the dog's "beauty" and "friendliness."

In their typology of dogs, South Punjabi dogfighters provide an elevated status to fighting dogs who, they say, represent their keeper's male status and honour. Fighting dogs are seen as being entirely different from other dogs in their behaviour and attitude, and believed to possess courage (*jur'at*), and an ability to outmanoeuvre the opponent through their strength (*zor*) and bravery (*dileri*). Thus, unlike any other dog, they are conceived as being worthy of human company, love, and care. Even though as a *haram* animal, they live in the proximity of dogfighters and inspire their *shauq*. When I asked dog-fighter Ajmal how he was able to develop his *shauq* with a *haram* animal, he responded by saying that fighting dogs were different, "they are an intelligent being," who "study" humans very closely. "Look," he said after making his dog shake hands with him, "they can even say Salam" (a gesture of greeting among Muslims). Munawar Chaklyar, another dogfighter, told me that *shauq* is never about *halal* or *haram*. Some people, he argued, have *shauq* of keeping *halal* animals such as cattle, roosters, quails, and partridges while others like the company of *haram* animals like dogs, bears, or monkeys. "This is all about *shauq* and personal preference (*apni apni pasand*)."

The *shauq* of keeping fighting dogs allows dogfighters to develop intimate bodily knowledge of canines through routines of affective labour and care. They spend days, months, even years together to understand and rec-ognise each other's preferences and priorities. This "affective proximity" (Govindrajan 2018, 80) enables dogfighters to develop more-than-human sociality and to fulfil their passion to train their cherished dogs for a symbol-ically rich fight.

The *shauq* of keeping a fighter

When I first visited Munawar Chaklyar, a dishevelled man in his early 50s, he looked at his thirty-one-inch high, large-headed dog with admiration and stated, "keeping and fighting dogs is my *shauq*." Nicknamed Manna, he was a small farmer and grew sunflower and cotton on rented farmland near his house. He also kept two cows and six buffaloes, but his fighting dog was the centre of his attention. As we discussed further, I learned that keeping and fighting dogs was something even loftier to him than mere *shauq*; it was *muḥabbat* (literally love or the activity you adore). He cupped the mouth-piece of a hookah and, after taking a long drag, explained this love, "There are no rewards or betting, we participate in dogfighting because it is our

Figure 4.4 Ajmal showing that fighting dogs can say Salam (shake hands).

muḥabbat." With time, this expression became clearer to me as dogfighters often addressed each other as *muḥabbattī* (a lover) and *shikārī* (a hunter). The terms used are revealing. The first one denotes an unmediated experience of devotion and love. The reason for using the second expression (*shikārī*), has more to do with the historic association of dogs with the hunt. Hunting was a popular pastime for the British during the time of the Raj (Pandian 2001; Hussain 2010). Later, the activity was adopted by rural landlords who accompanied dogs to retrieve killed game. As stated earlier, only affluent and politically influential men carry out illegal hunting activities in South Punjab, since they can obtain a hunting permit by using their connections with the authorities. Arguably then, dogfighters use the term *shikārī* when addressing their fellow men to symbolically give them higher status and respect. Another possible reason for using the term *shikārī* is the overlapping roles of dogs in the hunt and in the dogfight. On both occasions, a man trains his dog for a quest, travels some distance with him in a vehicle (jeep or light truck) to the site, and expects his dog to succeed and bring a reward to his keeper (whether by winning a dogfight or retrieving hunted game). Both terms mark a community of belonging that is inclusive and unique.

To belong to this community, a person needs to make sacrifices, both social and economic. Dogfighting is a very expensive *shauq* (*mhāngā shauq*), and not all people can afford it. Manna explained that a dog commonly consumes milk, butter, *ghī* (clarified butter), and meat in almost equal quantity to a wrestler's daily dietary needs. Although Manna could afford to feed his dog, for some people, keeping a dog meant serious financial sacrifice. Amin Uttera, a tailor and an avid dogfighter from village Havilian Naseer Khan claimed that "preparing a dog for fighting is either a *shauq* of an *amīr* (a rich man's enthusiasm) or a *faqīr* (a beggar's enthusiasm)." Nicknamed Meena Darzi (or Meena the tailor) by his friends, he had kept fighting dogs since childhood, and realised that the *shauq* was more expensive than raising an expensive, good breed, bull. "An affluent person," Meena explained, "can easily feed his dog, while a beggar goes from street to street and begs, usually getting more wheat flour, local bread (*rotī*), milk, and other things that are surplus to his needs, so he can give them to his dog."

Meena managed his *shauq* by stitching garments and sometimes working as a casual labourer on farms, and made $125–150 a month. As he spent a major portion of his income feeding an expensive diet to his dog, to see him grow healthy and strong, he was able to compete and challenge other dogfighters of his village. This investment of money despite poor circumstances, along with affective labour and enormous time, has turned dogfighting into a classless activity and made it accessible to people from different socioeconomic backgrounds.

The protein-rich diet for a fighting dog is selected cautiously and prepared with care. The ingredients are the same as the diet of a human wrestler, with butter, *ghī*, milk, almonds, and meat. Raw meat and *ghī* are considered important for increasing general physical strength, while *rotī* (local wheat bread) is deemed crucial for bodily growth. Ashiq, a dogfighter and the father of a wrestler, argued that when *rotī* is prepared with local method by kneading the wheat flour with salt and water and then cooking it in the clay oven to later coat it with butter, it becomes a vital source of growth for humans and animals. Ashiq grows cotton and wheat on his small agricultural farm just across the river Sutlej, and manages to store two sacks of wheat (each containing 100 kilos) for his fighting dog. He regularly purchases goat intestine from the town's butcher, cooks it in black pepper and salt, and adds a quarter kilograms of *ghī* before giving it to his dog to help increase his courage. A few weeks before a fight, Ashiq feeds apple jam (*murabbah*), cooked meat (with as much care as if it were being cooked for humans), *thada'ī* (almonds mashed in a mortar and pestle and diluted with milk), *lassī* (a yogurt drink), butter, unboiled milk, and *ghī* to his dog. This carefully selected dietary regimen, he says, turns a good breed dog into a ferocious fighter. However, the key is purity. The bread, butter, *thada'ī*, *lassī*, and most importantly, the *ghī* should be pure to be effective. "Dogs that are raised on pure (*khalis*) *ghī* and pure hard work (*khalis mehnat*) can fight with valour in the arena," he says. For him, pure *ghī* is vital and cannot be substituted by "Dlada" (a hydrogenated oil): the latter replaced *ghī* in South Punjabi villages in the past two decades. For many dogfighters of Ashiq's generation, homemade *ghī* is still the source of power and purity while market-based Dalda is seen as artificial and less-powerful.[1]

Figure 4.5 Manna with his dog, Tanger.

Whether it is the churning of the milk into butter, a distillation of the butter into *ghī*, or blending yogurt with milk to make *lassī*, it is usually milk that holds central place in a fighting dogs' dietary regimen. Milk (sometimes boiled, but mostly raw) is considered the source of physical strength and vitality, which in the form of *ghī* generates stamina and in the form of butter and *lassī* develops muscles. Such diets are expensive to maintain. Dairy products and other dietary items given to a fighting dog can cost up to 1000 Rupees (about $6) daily, however, most dogfighters keep cows specifically to provide easy access to milk products for their dogs. Such preferential treatment to fighting dogs, as I discuss in the next chapter, sometimes results in domestic disputes since the wives of dogfighters complain that their husband's *shauq* is prioritised over the needs of their children.

Figure 4.6 A dog keeper feeding his dog.

The regular provision of high quality dairy products and meat requires a serious investment of time, energies, and money, which is possible when a strong motivating force is behind it. *Shauq* is that motivating force for many dogfighters that enables them to develop an affective labour with their canines and makes them decorate the *haram* animal with expensive leather collars (*paṭā* or *paiṭī*) and various kinds of leashes. Some dogfighters even colour the dog's coat with henna, drawing patterns and sometimes writing the canine's name, while others prefer to crop their dog's ears to enhance his ferocious look. Cropped ears are also useful during combat because the rival dog does not get an opportunity to grip them in his teeth. Yet, a good diet or decoration is unimportant if a dog cannot fight well in the arena. The capacity to fight in a dog, it is believed, comes from birth and is not something that can be developed through diet or training.

Figure 4.7 Cutting the ears and colouring the body with henna are common decora-
tive practices of South Punjabi dogfighters. This kind of marking is also
indicative of wild animal traits, particularly the dark spotted pattern and
the eel stripe down the back. Most dogfighters have an underlying desire to
make the dogs look wild.

To breed a fierce fighter, a good quality pair is required. Unlike cockfighters,
dogfighters do not try to preserve the breed; rather, they find a champion male
dog and a female dog of good height, and outcross them. It is a popular belief
among dogfighters that a victorious dog will breed prize-winning pups, just as
a champion wrestler's son will become a successful wrestler. The quality, it is
believed, lies within the semen of the male dog which is considered the locus of
all strength and power.[2] Fighting dogs are thus kept under a strict regime of
celibacy: they are not allowed to go near a female dog for the duration of their
fighting life (dogs usually fight from two to eight years). However, successful
fighting dogs are allowed and even encouraged to mate once they finish their
career to produce excellent progeny. The common breeds in South Punjabi dogfighting are Bulli Kuttā, Jhogi Kuttā, Naṇgī Kuttā, Tulhy, and Gull Dongs, and
almost all of them are outcrossed with the aim of producing a tall, ferocious dog.

Once the pups are born, they are scrutinised and vaccinated by dogfighters.
Most of the time, the healthiest and most energetic male puppy is kept and
others are sold or shared with friends. The chosen puppy is then provided
with a vaccine, usually Hexadog (locally known as "*France ala tekā*" or injec-
tion imported from France), to keep him safe from rabies (*chitā-pan* or *halkā-
pan*), leptospirosis (*phetā*), and distemper (*khang*). The vaccine costs up to $6,
a huge amount to many poor dogfighters, yet almost everyone wants to have
it administered. Such hard choices reflect a dogfighter's sustained investment
in the animal who wins him *izzat* in the fighting arena. It also suggests that
an animal that is considered *haram*, culturally impure, and usually neglected
by the vast majority of people, succeeds in developing close relations with his
human keepers, albeit in an ethically complex way.

After vaccination, an important *rite de passage* is for the dog to be given a personalised name that turns him into an individual. A distinct name makes the dog a part of the dogfighter's social life, and sometimes also depicts characteristics that the dog is expected to possess. For instance, the name of Manna's dog was Tanger (Seraiki pronunciation of Tiger), because he hoped his dog would fight like a fearless tiger. Makhan's dogs were named Rustam (a mythical character in the Persian epic, *Shahnameh*) and One-Two-Five (referencing a powerful Honda Atlas 125cc motorbike). Some dogs are called after legendary fighters of the screen, for example, Bruce Li and Rambo. Other names followed certain incidents, for example, Riaz, a dogfighter and a cattle herder in his early 30s, named his dog Pistol because he had sold his pistol to buy the dog. Female dogs are only used for breeding and given "feminine" names that demonstrate their value and beauty, such as Bilo (female cat) and Motī (Pearl). Providing distinct names to dogs raises the social status of dogfighters in the village and reasserts their passion. For instance, when people identified Basheer in village streets as "the owner of Bruce Li," they not only recounted the successes of his dog in fighting competitions but also acknowledged Basheer's monetary investment, hard work, and genuine *shauq*. After birth, vaccination and naming are the two important practices that make a dog ready to be trained for a fight where high symbolic stakes, such as winning or losing *izzat* in front of a large audience, in particular before a person's jealous relatives, are involved.

Along with a "powerful" diet, the fighting dog also goes through rigorous physical training to increase his strength and stamina. Usually, this involves long-distance running (*tora*) and occasional sparring exercises (*tapānah*) with other dogs. I sometimes accompanied Riaz when he took his dog running with a friend on a motorbike. The friend or relative would drive slowly over soft earth for about ten kilometres while Riaz sat on the rear seat, holding the leash. This was followed by a careful, dry massage to relax the dog's muscles and develop a strong bond with the keeper. However, this bond is tested closer to the dogfighting season when the dog is sparred to sharpen his skills. Dogfighters say that not all dogs fight in a similar manner, some prefer to lock their teeth on the shanks of the opponent while others bite on the neck; some grab any part of the opponent and hold tight, while others always look for an advantageous position; some can breathe while biting while others get exhausted easily and cannot keep their hold. After sparring a dog against those with a different fighting style, the dogfighter tries to develop a number of coping strategies in the canine.

The training regimen of a fighting dog resembles that of a local wrestler. They both go through stamina and strength exercises, that are followed by a daily dry massage.[3] Both are also encouraged to sleep long hours, sometimes more than eight hours a day, for optimum results. Yet, some dogfighters contend that training a dog takes more effort than training a wrestler since the latter is self-motivated whereas a dog requires the assistance of his keeper at all times. As we will see below, the comparison between dogs and wrestlers

Figure 4.8 Makhan's dog returning from his daily run.

also holds good in the fighting arena, because the success of the dog in a fight is not simply a victory for his keeper but also means a triumph for all friends, relatives, and supporters. I suggest that the sole purpose of organising a dogfight is to reproduce hegemonic traits of masculinity by outwitting the opponent in front of a large audience, thus achieving *izzat* through the manly performance of one's dog who is capable of emasculating the opponent.

The fight of honour

Both dogfighting and wrestling are held as part of local *melas* (village festivals) for the entertainment of rural men. The word *mela* originates from the Sanskrit word *mel*, connoting a religious fair where a large number of people congregate (Kumar 1988). In South Punjab, *melas* are usually organised in springtime to provide a holiday after the wheat harvesting season ends. Most are held near a shrine to also celebrate the anniversary of the saint. They are important cultural events, full of recreational opportunities that involve economic exchange (see Chakrabarty 1992, 543). As people congregate to purchase goods or to eat food, they also get the opportunity to visit *mela* bazaar and buy things such as toys, trinkets, clay pots, metal and wooden vessels, and other inexpensive knick-knacks. Elsewhere (Kavesh 2018a), I have discussed in detail the politics behind the organisation of South Punjabi *melas*, examining the history of colonial manoeuvres to ban large gatherings of people in the fear that this may turn against the Raj. The more recent adoption of such

Figure 4.9 A sweet shop at the mela bazaar.

prohibitions by present-day rural politicians and district bureaucrats serves the same purpose: to retain hegemony in the area.

In the mid-1990s, I remember, *melas* were frequented by snake charmers, acrobats, and monkey trainers who would captivate the crowd with their performance. In my school days, I visited *melas* regularly to enjoy local sweets and amusement rides. There were dancing bears, gypsies, and traditional healers who would sell inexpensive herbs to the largely male audience. Today, organisation of *melas* are more structured under the control of the state through district bureaucrats. There are no snake charmers or gypsies, although gambling and other illegal activities take place with the informed consent of the police and district management authorities. The crucial feature of these *melas* today are spectacles like wrestling, horse dancing, bullock racing, camel fighting, and dogfighting, organised in the open grounds near the *mela* bazaar.

A *mela* is a festive place, a liminal space and time when the *shauq* of dogfighting may be performed before thousands of people. The annual dogfighting events generally take place on the second and busiest day of the *mela* when the organisers ask experienced dogfighters to manage the event and attract a crowd. Sometimes, transportation is provided by the organisers to participating dogfighters and their canines but there are no monetary rewards for the winner. Unlike cockfighting, the practice of fighting dogs does not involve a wager, either from the organisers, the dogfighters, or the crowd. Spectators congregate to praise the "true" masculine qualities of the canines and their

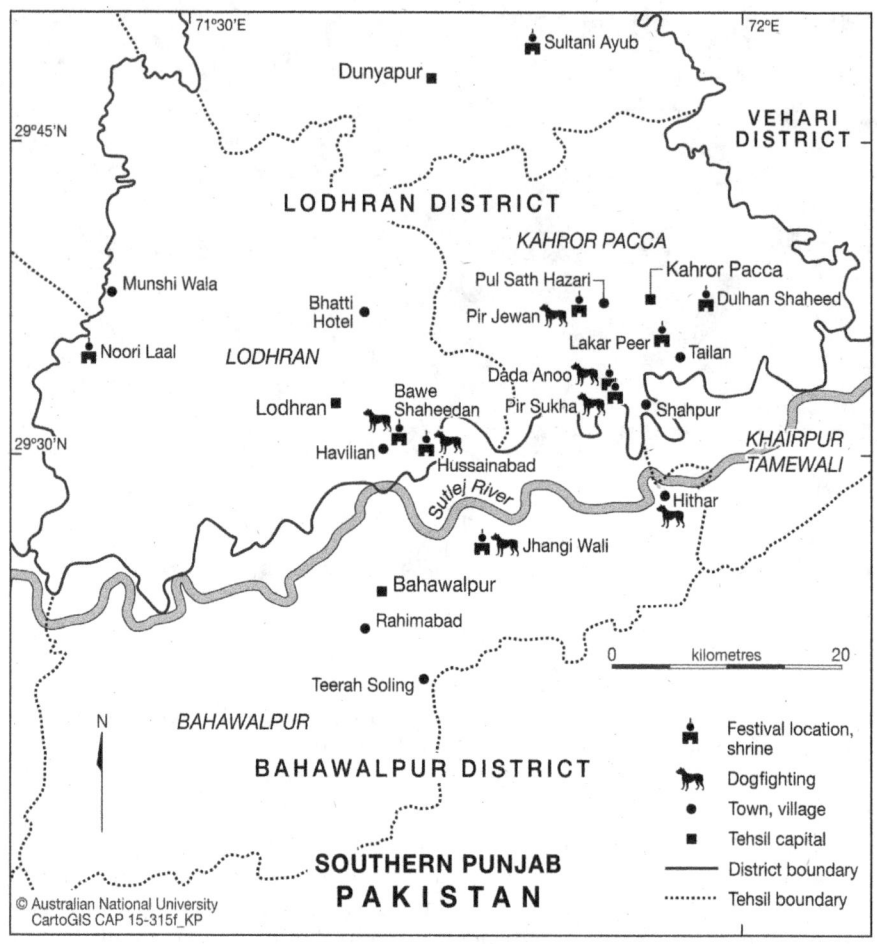

Figure 4.10 Map showing festivals and dogfighting in different areas of South Punjab, Pakistan.

keepers. Even though monetary rewards are absent, competition is fierce. The ability to bring forward a dog that commands the popular traits of courage, bravery, and strength is what's at stake. When victorious, these traits along with prestige and honour, flow towards the dog's keeper and his supporters.

Dogfighting (*kuttyāṇ dī malheṇ*) is usually held in the morning around ten and lasts until the afternoon. There are no intervals or rounds and each fight usually lasts between twelve and fifteen minutes. The almost two-acre wide arena is circled by a large crowd of men who arrive on motorbikes, bicycles, rickshaws, cars, and tractor-trolleys to watch the bouts. Fighting pairs are announced on loudspeakers placed at opposite ends of the grounds.

Figure 4.11 A dogfight in progress.

The announcer receives a small amount of money from both organisers at the end of the day (1500 Rupees from each organiser, about $18 in total). Local minstrels, such as drummers (*ḍhol waly*) and shehnai players (*bein waly*) are always present to add colour to the celebrations and earn money from the winning group for their services.

After all the pairs are decided, the first fighting pair is washed to remove any noxious substances. Washing is a crucial component of dogfighting because, unlike pigeon flying and cockfighting, the support networks behind a fighting dog are far more extensive and this raises the stakes of the competition and increases incidents of cheating. For instance, Riaz told me, one way of gaining advantage over the opponent is to rub sheep's urine on the dog's coat prior to the fight. This arouses the sexual senses of the rival dog and makes him lose focus on the attack. Some opponents rub toxic substances on to their dog's body, thus putting the opponent dog at danger of being infected and poisoned. To avoid such incidents of foul play, two steel tubs full of water are kept in the arena and the keepers pour water over their dogs and rub their bodies before unleashing them. This cleansing of the canines' body before each dogfight assures the audience, and all concerned parties, that this combat over *izzat* will take place on fair terms.

As the canines are unleashed for combat, both animals lunge at one another with ferocity, locking razor-sharp teeth around each other's bodies, holding on with all their might, and battling to gain supremacy. Blood flows,

Figure 4.12 A dogfight in progress.

tails are raised, and the animals fight relentlessly. The crowd of thousands of men watch the battle with frenzy, curiosity, and amusement. At times, the spectators fall completely silent, focusing on the attacking and defensive moves of the dogs. This silence, as noticed by Alter in the context of North Indian men's wrestling, may be read as a marker of true appreciation:

> once a contest has begun everyone becomes quiet and attentive. This is less a natural reaction than an issue of decorum. Silence serves to focus attention on the wrestlers, and the quality of a bout is said to be reflected in the degree to which skill and strength can leave one quite literally speechless. Periods of silence are counterpoised with eruptions of vocal empathy. But silence is the mark of true appreciation.
>
> (1992, 180)

To decide the winner in a dogfight, two referees walk around the dogfighting pit and listen intently to any sounds the dogs make during the bout. Their authority is absolute and unchallenged. The fight ends when one of the dogs barks (*boṇknā*), yelps (*ḍaṇḍī kaḍhnā*), growls (*yakhi mārnā*), or turns away from the rival (*pāsā mārnā*). Dogfights also end when the keeper of the dog concedes and calls the fight off (*hath kharā karnā*). If both dogs keep fighting for more than fifteen minutes, the fight is usually declared a draw with the consent of both keepers.

Dogs rarely die in South Punjabi dogfights but I saw many of them getting severely wounded. When the beautiful creamy coat of a fighting dog became red with blood running from his leg, while his keeper kept pushing him to fight, it seemed to me that the attachment that developed between a man and his dog over long hours of breeding, vaccination, naming, feeding, and even adorning the canine was less meaningful for the dogfighter than the symbolic reward of the dogfight. The *haram* animal from whom most people kept a distance in their everyday life was becoming worthless to dogfighters too in the fighting arena. But, perhaps it was the liminal space of the *mela* that converted the attachment between a man and his dog into detachment, and expected from both to display hegemonic traits of masculinity through their behaviour and practice. There thousands of people expected the dogfighter to maintain his composure during the fight and display self-control to retain his *izzat* while the canine was expected to take it "like a man" when being whipped by the opponent dog, and to not give up.

When the fight ends, the wounds of the dogs are treated with herbal ointments. After twenty to thirty days, when the victorious dog recovers, he may enter another fight. The losing dog (*kaṭā huā kuttā*) is generally sold as a guard dog because dogfighters contend that once a dog loses a fight, he becomes habituated to losing (*ādī*) and there is little option but to sell him. This is very different from other types of *shauq* discussed in this book. While a pigeon may be captured by a rival fancier or a cock may die, in dogfighting,

Figure 4.13 Dogs are constantly pushed to fight despite sustaining severe injuries.

the loser dog suffers "social death." He is not allowed to fight in another competition for life and, in most cases, is divested of his previous privileges. Such treatment is criticised by pigeon flyers and cockfighters who accuse dogfighters of not possessing "genuine *shauq*." Although the defeated dog is not killed in South Punjab, as opposed to in other parts of the world, the dog loses everything, including quality diet and regular massage, and is abandoned by the dogfighter and his supporters.[4]

Since there is no betting involved, the *shauq* of dogfighting has been described as a non-material activity where the entire struggle is calibrated for prestige and honour (*izzat*). This is, however, in line with Bourdieu's (1990) notion of symbolic capital which serves as a type of currency that can potentially be converted into material profit or economic capital. He argues, "[t]he interest at stake in the conducts of honour is one for which economism has no name and which has to be called symbolic, although it is such as to inspire actions that are very directly material" (1990, 120–121). Looking at the symbolic structure of Algerian society, Bourdieu asks us to grasp the "economic rationality of conduct" even if it appears to be symbolic in nature (1990, 120). In South Punjab, the economic rationality of dogfighting can be grasped by taking into account the structure of the activity. For instance, the decision to spend an enormous amount of money and effort to ready a dog for the fight seems irrational and economically aberrant considering that there is no monetary reward for the winner. Arguably, dogfighters justify this spending spree with reference to a culturally meaningful term—*shauq*. I do not doubt their explanation or their passion, but there are some obvious pecuniary benefits. If the dog wins, his value increases manyfold, and the keeper can make a significant amount of money by selling him or selling his pups. In other words, the whole process can be described as cross-cutting three spheres: (a) the dogfighter spends money and resources to ready his dog for the fight, (b) the dog wins a combat and symbolic capital is channelled to his keeper, and (c) this symbolic capital has the potential to be converted into economic capital through breeding and selling the pups. Thus, the monetary benefit that a dog's keeper derives is related to the dog's ability to win, leading to the accumulation of symbolic capital (which Bourdieu termed "credit of renown" [1990, 120]). Equally, it is important to note that people do not consciously calculate material benefit from symbolic gains but, as Bourdieu states, different forms of capital have differing saliency across fields, and in some circumstances, they overlap and can be converted (1990, 123). This is partly why the activity of dogfighting continues to be identified as a platform for only accumulating symbolic capital rather than acquiring any economic benefits.

The joy of defeating rivals

The performance of dogfighting, in the sense in which I am using the term here, is not only about showcasing hegemonic masculine characteristics but also provides opportunities to South Punjabi men to be recognised and to gain social status. For example, Riaz claims that wherever he goes to graze his cattle, people recognise him as Riaz *kutty walā* (Riaz, the dog keeper), "Everyone knows me in a 200-kilometres radius in every direction because of my dog." Winning a dogfight also helps win friendships with other dogfighters who share the same passion and understand each other's *shauq*. Basheer Laar reflects that when he fought his dog in Hasilpur (250 kilometres from his home), many dogfighters approached him after the fight and later became good friends with him. "I did not know anyone there, but soon as my dog won the fight, many people asked for my mobile number. Some later called and even visited me." This friendship between *shauqeen*, which Evans and Forsyth (1998, 55–60) critically describe as bonding among the fraternity of dogmen, helps the men discuss and develop new methods of breeding, feeding, and training their canines.

The "true" test of friendship takes place in the dogfighting arena, when friends are differentiated from *shareek*. The term *shareek* literally means those who share land or inheritance and in Pakistan, it is generally used for agnatic rivalry, mostly to indicate hostility with patrilineal cousins (see Lyon 2004, 127; see also Barth 1959, 65–66; and Ahmed 1980, 3). In everyday usage in South Punjab, *shareek* means jealous relatives or friends who compete for *izzat*. People strive to outwit their *shareek* in every sphere of life such as in having a better house, a better harvest, better quality livestock, or marrying into rich families. Since there is always implicit hostility between *shareek*, the dogfighting arena becomes a crucial occasion when these rivals can be humiliated in front of a large audience. The joy of fighting dogs, they say, lies less in the victory and more in defeating *shareek*. Celebrations after winning a fight, from dancing to placing a rose garland around the winner person's neck is directed specifically towards defeated rivals and their supporters. For this reason, friends who do not participate in the winning celebration are seen as envious and disloyal. Explaining the value of sincerity amongst friends, Meena Darzi argues, "only *shareek* stay back during the celebrations, a true *yār* (friend) never feels shy." With friends, dogfighters say, a man enjoys the pleasure of victory. This is often captured in a famous Seraiki language saying: "*yārān nāl bahārān*," meaning, with friends there is spring (or one can be truly happy only in the company of one's friends) [see Narayan and Kavesh 2019].

The South Punjabi *shauq* of fighting dogs, like Michael Herzfeld's Cretan mountain villager's practice of sheep theft (1985), creates new friends and allies and serves an immediate notice on *shareeks* of the dogfighter's skill—his ability to breed, feed, and train his dog as a ferocious fighter. Sometimes, a loser may blame his *shareek* for cheating and harming his dog through the evil eye. I will discuss the threat of the evil eye to *shauq* in Chapter 6, however, here I emphasise that to South Punjabi rural men, dogfighting is less about the physical combat of canines and more of social performance to humiliate *shareek* in front of a large audience. Winning a dogfight at a *mela* triggers the crowd's admiration for presenting idealised masculine values through

Figure 4.14 True friends are recognised by their eager participation in winning celebrations.

a canine's fight and provides the keeper with higher *izzat* than the *shareek*. Each dogfight is important; every success marks progress in that symbolic struggle, and every loss takes a person away from being recognised in the village environment.

In the symbolic struggle of men to gain ascendency, the dog remains a key player. Despite severe injuries, he is valued for displaying absolute courage and unmatched bravery. In the arena, the non-human becomes evidence of his human keeper's efforts, an embodiment of his *izzat*, an emblem of his *mardāneat* (masculinity), and a goal of his expectations. In this social performance, the dog, like a village wrestler, is praised for welcoming all challenges, and enabling his keeper to hold his head high with pride and dignity.

Overall, dogfighting in South Punjab is a complex *shauq* that is performed as a part of festivals to achieve *izzat* and display traits of hegemonic masculinity. It is complex first, because it requires that a man develop a close relationship through affective labour with a *haram* animal despite strong religious disapproval. Second, this *shauq* is complex because fighting dogs does not involve any betting and is staged to achieve symbolic rewards in front of thousands of villagers by outwitting the opponent. Interestingly, similar symbolic rewards are sought by the educated, urban, middle-class population of central Punjab who would, in a general sense, strongly oppose dog fighting as a barbarian practice of remote villagers. A good example is dog shows organised by the Kennel Club of Pakistan in different urban centres on an annual basis.[5] Sponsored by popular brands of pet food (such as Royal

Figure 4.15 Dog, the animal that ensures a man's *izzat*.

Canin), these dog shows judge the purebred and reward the dog keeper with internationally recognised certificates. However, a number of factors turn this show into a display of wealth and power and away from the qualities of a dog. For instance, there are many foreign breeds of dogs like Huskies, Pointers, Poodles, Labradors, or Cocker Spaniels, some of whom finds it extremely difficult to cope with the hot and dry climate of Pakistan. Specialists are then hired to give care to these foreign breeds and specialised servants are kept to manage their daily needs like feeding them and taking them on daily walks. This makes the exercise a display of conspicuous consumption, staged to achieve symbolic reward rather than emotional benefit (Veblen 1912). Because of substantial media coverage, these dog shows function as a sign of distinction for rich dog keepers (Bourdieu 1984).[6] In this modern *shauq*, canines are not staged for fighting but are raised to accumulate status in other socially important ways. They are not expected to display courage, bravery, and strength but are praised for their friendly behaviour towards other dogs and humans, and for their composure and flexibility in being handled by the judges. In this context, where evaluation is based on aesthetic standards, the value system shifts to spotlight the class difference.

In this chapter, I have tried to explicate how the dog, as a polluted animal, affects the social life of rural South Punjabi people in various ways. For dogfighters, the fighting dog is a part of their *shauq* who is incorporated into their daily lives, kept with great care, given personalised names, and provided a quality diet. In the fighting arena, the dog, like a human wrestler, is expected to display courage (*jur'at*), bravery (*dilerī*), and strength (*zor*) in front of a large crowd and help the keeper achieve *izzat* and humiliate his jealous rivals. The winning dog can turn the symbolic gain of his keeper into material profit if he is kept for breeding after a successful fighting career. A losing dog, however, loses all previous privileges of care and attention and is sold as a guard dog. The keeper of the defeated dog tries to buy or breed another canine and continues to practise his *shauq* until he re-emerges victorious and reclaims his honour.

As a "favourite pet" of the twenty-first century, the dog has a special place in human societies. Many people see the dog as a companion whose relationship to the keeper is long-lasting, marked by loyalty and unconditional love. Reports of cruelty to dogs thus receive considerable criticism from animal rights groups. For instance, when National Football League superstar quarterback, Michael Vick, pleaded guilty to the practice of dogfighting in August 2007, the US media and general public criticised this "brutal sport" as "a reprehensible act" and he was condemned harshly by animal rights activists (Laucella 2010; Coleman 2009). However, dogfighting in rural South Punjab, like bullfighting in Spain, or greyhound racing in Australia remain culturally important. And yet this *shauq*, along with pigeon flying and cockfighting, generates its critique from those who interact with the *shauqeen* on a daily basis, including the animal keepers' wives and other family members. In the following chapter, I explore how animals mediate a relationship between the *shauqeen* and their wives in a multi-species household.

Notes

1 Similar observations were noted by Alter (2002, 86). See also Kumar (1988).
2 While writing about wrestlers of North India, Joseph Alter found similar beliefs regarding the power of semen (see Alter 1992, 129, 149).
3 Body massage in Pakistani Punjab is a fine-tuned technique with pressure, vibration, and friction (see Frembgen 2008, 7). For both the wrestler and the dog, massage helps in developing strength and suppleness.
4 See, for example, Evans and Forsyth (1997), Evans, et al. (1998), Kalof and Taylor (2007), Kim (2015), Coleman (2009), Homan (1999), Smith (2011), Laucella (2010), Iliopoulou and Rosenbaum (2013).
5 These dog shows are organised in association with Fédération Cynologique Internationale (FCI), a Belgium-based representative organisation of Kennel Clubs from ninety-one countries. The organisation aims to promote, set standards, and keep records of purebred dogs.
6 From 2008 onwards, major English-language newspapers (*Express Tribune, The News*, and *Dawn*) and television news channels (Geo News and Dunya News) started substantial coverage of these dog shows and are reshaping the discourse on the human–dog relationship in Pakistan.

5   A life with *shauqeen*

Familial relations in a multi-species household

In a multi-species household, pigeon flyers, cockfighters, and dogfighters form a kinship of needs and responsibilities, care and affection, and attention and attachment between their animals and family members. The men keep their cherished non-human companions in or near the household, provide them with such care as they provide to kin, devote time and energies on them, and make them share the space with family members. However, such sharing leads to new expectations that, as Ben Campbell (2005, 80) notes while discussing animal husbandry in Nepal, foster entangled bonds of intimacy. These bonds are entangled as, in most cases, the men end up bestowing more care on their animals because they hold the promise of honour, glory, and prestige in front of peers. Perhaps the most vulnerable to such preferential treatment of animals over the family is the animal keeper's wife. In this chapter, therefore, I pay close attention to how the wives of animal keepers view their husbands' *shauq* and examine men's strategies and efforts in (re)shaping and (re)forming their life priorities. A focus on the wives of animal keepers is important also because it provides an alternate view on men's choices and practices.

In South Punjab, most women remain uneducated, and this was the case with the wives of animal keepers who were interviewed. The age of these women ranged between 30 and 50 years and a majority of them were married in their early 20s with an average of one to four children. I was able to speak directly with some of them and engaged in protracted conversations with my own female family, friends, and acquaintances. However, in Muslim South Punjabi society, meeting and interviewing women goes against religious and social norms, so I employed two female Research Assistants (RAs) to help me get in-depth, qualitative interviews of women relatives of the animal keepers. They recorded these two to three hour-long interviews and I transcribed the recordings with the coordination of the RAs. Whenever the transcription raised further questions, the RAs visited the village women again and asked for missing information.

As I share women's stories and illuminate their daily living arrangements with animal keepers, I explore "what stories make worlds, what worlds make stories" (Haraway 2016, 12). My examination of the women's stories and their lifeworlds with *shauqeen* helps me carefully study the various effects of a man's *shauq* on his family members in a multi-species household. A multi-species household, as sociologist Nickie Charles (2014) uses the term, refers to the place where the animals live in intimate proximity and form sociality with humans, such as the rooftops of pigeon flyers, and the courtyards of cockfighters and dogfighters. Since the lives of pigeon flyers, cockfighters, and dogfighters are entangled with their non-human companions, I explore the affects of this entanglement on the wives of the animal keepers and examine how the women see their husbands' attachment to their animals as something that hinders the performance of their manly responsibilities towards the family. By understanding this inter-species relatedness through the women's life stories, I delve into the women's own interpretation of honour and masculinity and explore the ways in which it affects their daily life.

A gendered understanding of masculinity helps us reinterpret the meanings of duty, passion, and honour in rural South Punjab. A "true" male, according to men, reproduces vital traits of hegemonic masculinity and gains honour by distancing himself from behavioural traits generally associated with women (Connell and Messerschmidt 2005). To many South Punjabi women, however, a true man is responsible for his family's well-being, and protects, cares and provides for them. This conceptualisation of masculinity lays emphasis on sociocultural norms and traits that are different from those discussed in previous chapters. An ideal male, women argue, defends family honour, avoids wasteful pursuits, uses time in a "productive" way, and prioritises the care of the family over his *shauq*. I explore these interpretations of masculinity below and show that when the men try to achieve honour by performing dominant traits of hegemonic masculinity in the village setting, they end up in a conflicted relationship of care and neglect within a multi-species household.

A relationship of care and neglect

The institution of the family in South Punjab, although in flux, remains relatively static. The family or *khāndān* in Punjab could be as small as a nuclear family (comprising only husband, wife, and children), or as large as a joint family (comprising many generations living together in the same household). In either case, the *shauq* of a man—an ingrained predilection that requires time, attention, and energy—has a deep impact on his relationship with close family members. As the majority of animal keepers I met during my fieldwork were married, this made their *shauq* a serious affair, something that the wives of animal keepers saw as a threat to the family unit.

The majority of marriages among animal keepers, according to the prevailing local custom, were patrilineal and arranged by the heads of families (usually fathers, grandparents, or paternal uncles). This means that men who were known to cultivate a *shauq* of animal keeping (which society frowns upon, see next chapter) were marriageable largely because marriages were endogamous in nature and intended to maintain family alliances. These arranged marriages among relatives significantly decrease the rate of divorce and make it a social stigma for both women and men. However, divorce is not rare in the South Punjabi context, and when marriage is not arranged between relatives, divorce appeared in its fiercest form. Take the case of Naseem who faced this reality more than once.

Naseem was born into a poor family and when her father arranged her marriage to a motorcycle mechanic, he told her that he was sending her off and expected only her dead body to return home (*teḍī ḍolī bhjeṇdā pyā haṇ, teḍā janazā wapas awe*). The eldest among three sisters and two brothers, Naseem was expected to set an example to her siblings by remaining with her in-laws for life. However, after only three months, her husband who was a heroin addict, divorced her over a small matter of poorly cooked food, shattering her life by uttering the word *talaq* (divorce) three times.[1] When she returned to her father's house, she felt like a "burden" (*bār*) on everyone. Only ten days before her *iddat* was to come to an end (the Islamic period of waiting after divorce or the death of a spouse before a woman may remarry) that her father accepted 150,000 Rupees ($950) as bride price (*ṣubhā*) from Bateel, a man in his 60s. "He was about 40 years older to me; older than my father. He asked me to live with him in his room and promised to perform *nikāh* (formal marriage) after ten days." Naseem recounted.

However, before Bateel could perform *nikah*, Naseem started noticing the admiring glances of Shan, the young nephew of Bateel, who talked to her for hours and made her laugh. The night before her *nikah,* Shan drugged the entire family by mixing sleeping pills in the dinner and eloped with Naseem to Karachi, where they had a civil or "court-marriage." Bateel was incensed and threatened to kill the young lovers if they ever returned to the village. Six months later, as Bateel's rage cooled, he swore on the Qur'ān to forgive his young nephew, saying that he was like a son to him. Naseem and Shan worked as domestic labour in a wealthy household in Karachi and only decided to return to Shan's village after persistent phone calls from Shan's mother who guaranteed their safety. Returning was easy for Shan but Naseem found that she had no *izzat* (honour) in Shan's household where her "merciless" mother-in-law, "clever" sisters-in-law, and an old uncle-in-law would address her as "bitch" (*kuttī*), "witch" (*bhoṇḍen*), and "wicked" (*badmāsh*), and loaded her with house chores in vengeance. She was asked to clean the house, cook for them, fetch drinking water from the tube well, wash everyone's clothes, and care for the goats. Even Shan seemed to have changed and did not care for her anymore. He started beating her at the least provocation and his mother would incite him to do this often till Naseem ended in serious bruises. When a relative demanded Shan repay the money that he had borrowed for the Karachi trip, Shan blamed Naseem for all the troubles in his life and, on his mother's advice, divorced her by tearing up the papers of the court marriage.

The day Naseem left for her father's house, she met Riaz on the bus station. She knew Riaz as Shan's friend, who was always kind (*mehrbān*) to her. Instead of going to her father's house, she opted to go with Riaz. This was the time when I began my fieldwork and knew Riaz as a dogfighter and a cattle herder.

However, in the village many men identified him as a man who kept Naseem in his house without *nikah*, only for sexual purpose. Riaz took some pleasure in this because he gained the reputation of being a *juoān mard*, a brave man, and an emblem of hegemonic masculinity. Naseem, however, was seen as a weak woman sold by her father in marriage, someone who was accustomed to changing husbands for sexual pleasure—a "*gashtī*" (prostitute). Naseem was aware of her reputation but, before she could do anything to alter her situation, she was pregnant with Riaz's child. Some months later, a daughter was born to them, destined to be an illicit (*harāmī*) child of a dishonoured mother. Now more than ever, Naseem wanted to change her status so her daughter could gain some *izzat* in the village. She went to the mother of one of my RAs, a widow respected in the village because of her late husband's political standing, and requested her to persuade Riaz's parents to convert this relationship into a *nikah*. Riaz's extended family agreed and they got married in a simple ceremony.

Riaz's *shauq* of dogfighting always held central place in this relationship. He would spend most of his day with his fighting dog, taking him for a run, massaging him, and preparing his feed. Meanwhile, Naseem's duties began early in the morning with cleaning the house, cooking for the extended family, washing the dishes, and accompanying her sister-in-law and mother-in-law to gather grass (*karī dā ghā*) for the cattle. This is a laborious task that requires "plucking pest plants with bare hands out of cotton and wheat fields, binding all of them together, and then carrying them home on the back" Naseem said. Upon her return, she was expected to feed her daughter, chop the grass manually, and spread it in front of cattle. Riaz would only turn up to milk the cows and kept the money after selling the milk. Naseem resented this because the most precious resource of the household was either sold to a milkman or kept for making butter and *ghī* for the dog. "I eat bread with salt and red chilli (*namak mirach de nāl*)," she said, "while the dog eats *ghi* every night." She remembered how sad she felt when Riaz gave one kilogram milk to his dog and not a single drop to their daughter, "I implored (*minnat kītum*) him to give half a glass of milk to our daughter before she went to bed, and after a month of pleading he listened to me."

For Riaz, feeding his dog was important. The canine was getting ready for the upcoming fight in a festival where thousands of people would judge his bravery and strength and praise Riaz for preparing a champion dog. The dog's success in the fight would mean that Riaz would gain *izzat* among peers that would raise his masculine status in the village. Spending time with his dog was a part of his *shauq*, and he told me when he massaged his dog's legs and body or prepared his feed, he felt like he had achieved something for himself. Naseem knew that the dog was being trained for an important fight but she condemned Riaz's overindulgence in this *shauq*: "He remains busy with his dog, even when I call him for his meals, he does not come." She found Riaz's *shauq* for fighting dogs as something that kept him away from his family, "every day after the massage, he places the *haram* dog in his lap like his child. I haven't seen him loving his daughter like this." This preferential treatment to the polluted animal, Naseem said, makes her think of her place in Riaz's life: "He has all the money for his dog but when I ask him for new clothes, he refuses." On the previous Eid, Naseem recounted, when everyone was buying clothes before the three-day-long holidays, Riaz asked her to go to the landlord's house to ask for some old clothes for herself and their daughter.

As I came close to finishing my fieldwork, I wondered what the future held for Naseem. She had led a troubled life and when she eventually came close to finding a relationship (a family, if we call it, with a husband and a daughter, and a house shared with extended relatives), the life she had was full of neglect, arduous labour, and emotional and physical abuse. The dog may not have been the reason for Naseem's troubles, however, the care he received from Riaz made Naseem compare her status to that of the canine. As Riaz prioritised care to his dog, Naseem witnessed a complex interplay of engagement and detachment within the household. Engagement and detachment, as Matei Candèa (2010, 254) suggests, are not always polar opposites, particularly in the entangled social relations between non-human and human selves, they are best treated together. In that multi-species household, the more Riaz engaged with his dog, the more Naseem read this *shauq* as the reason for his emotional detachment from her and their daughter. As Naseem encountered daily neglect by Riaz and observed his care and affection towards the dog, she saw the canine as an actor who actively influenced their life choices, behaviour, and practices.

When human and non-human lives unfold alongside each other, as Govindrajan (2018, 4–10) shows, an entangled relatedness develops, which is not always "nurturing or heartwarming" but can involve hatred, exclusion, and abuse. Although Riaz provided quality care to his fighting dog, Naseem had to ask for months on end for new clothes or beg for half a glass of milk for their daughter. In this multi-species kinship, Naseem experienced the neglect and detachment that relatedness sometimes involves.

Nasreen's life was influenced by the South Punjabi conceptualisation of "honour." I could have started her life story when she entered Riaz's life to keep my focus on how *shauq* affects a man's relationship with his family. However, I chose to start her narrative a few years earlier to show how the category of honour shapes people's lifeworlds in Muslim South Punjab. Pakistani feminists and scholars writing on the issue of honour (e.g., Brohi and Zia 2012; Jafri 2008; Khan 2006; Shah 2016; Zia 2019) have discussed in detail how this concept is a crucial navigator of women's lives and actively influences their choices, values of modesty, and social standing.

While rethinking this chapter, I called my Pakistani RA who told me excitedly about her encounter with Naseem's daughter and her sister-in-law. "How is Naseem?" I asked, "Her daughter says she died," my RA said. I was taken aback. "No, no, she did not die," my RA assured me after guessing my sadness, "she eloped with someone from Rohi and her sister-in-law says we have told her daughter that she died."

Finishing the call, I was happy for some inexplicable reason. Maybe it was because of Naseem's retaliation, her courage to break free of centuries-old customs that bound her to a loveless, arduous life and that continue to shackle many other South Punjabi women.

Embodying masculinity in everyday familial relations

As discussed earlier, masculinity (*mardāneat*) in South Punjab is a fluid concept that transforms itself in different social contexts. This leads us to consider the existence of multiple masculinities that coexist to shape the behaviour of men in their everyday social dealings. However, as Raewyn Connell and James Messerschmidt (2005, 832) suggest, in every culture there is a dominant form of masculinity that clings to the hegemonic position in the gender order and embodies the "most honoured way of being a man" by ideologically legitimising the subordination of women to men. It focuses on physical power, competitiveness, and emotional detachment (Bird 1996, 121), and although is not adopted by all men in all situations, it stands as "a reference point with which other forms of masculinity correspond" (Avieli 2012, 61). As discussed earlier, my conceptualisation of masculinity takes inspiration from Connell's (1995) explanation of hegemonic masculinity, particularly her suggestion to identify multiple masculinities in a society. In South Asia, masculinities appear in varied forms, such as passing time in limbo after college (Jeffrey 2010), idly spending time on culturally important spots like *addās* (Chakrabarty 1999), displaying subaltern resistance through manly demeanour (Rogers 2008), through everyday gendered relations (Chopra, Osella and Osella 2004), and sometimes by empowering the powerless through cinema (Mazumdar 2007). In all these situations, masculinity becomes an ideal, a societal expectation, and a value that demands men act, behave, and shape their gendered relations in specific ways.

In South Punjab, masculine expectations form a man's practice in public and private settings and influence his conceptualisation of honour. Early one October afternoon in 2008, when I started my fieldwork with pigeon flyers in a South Punjabi village, I accompanied Khudla to a sorcerer to remove the curse on his marriage, I knew it was his masculine status in the village that he was trying to defend. When Khudla reached his mid-40s, his desperation to find a wife multiplied and he tried all tricks: requesting his brothers to arrange his marriage, politely asking friends, fostering new contacts with remote villagers, and he even tried contacting some poor families who bluntly turned down his marriage proposal. It was perhaps his *shauq* of pigeon flying, his lower socioeconomic status in the village, or his age that made people refuse him their daughter. Some people laughed at him behind his back, others called him *bechara* (poor thing), and some people taunted him by asking about the progress on his marriage proposals. This affected Khudla's self-esteem and, most importantly, his masculine status in the village. Earlier, a friend had told him about a sorcerer (*āmil*), a Bengali baba, who had just arrived in a nearby town and guaranteed a marriage for single women or men in just 1000 Rupees. Khudla offered him the money and a living crow and a black chicken, which the sorcerer sacrificed to cast-off the dark spell. It had been a month-and-a-half since the sorcerer performed the ritual and there was still no sign of a marriage proposal. Khudla was getting impatient, so we decided to visit the sorcerer.

As we reached the sorcerer's abode, a two-room brick house with a dead tree in the centre of a large veranda, we carefully removed our sandals before entering the dark room filled with the fresh smell of burning incense sticks. The sorcerer, clad in a green shawl with steel bracelets around his wrists and agate rings in all fingers, was attending to other clients. As he concluded business with them, he wrote a *ta'veez* (a paper on which a spell is written) for Khudla and commanded him to bury it in front of his doorway, "a marriage proposal will soon come" he proclaimed. Within two weeks, Khudla's distant uncle offered him his daughter Kalsoom in marriage. Kalsoom had just turned twenty years of age at the time of her marriage. Her father, a farmer and a cattle herder, had five daughters while he waited for his wife to bear him a son. Kalsoom was the eldest among them. On his marriage day, I remember joyous Khudla invited the entire village and ordered the village chef to cook twelve cauldrons of meat pulao. Kalsoom added to his joy by bringing a "good dowry" (*changā ḍāj*), including two charpoys, a five-foot-long trunk filled with quilts, bedsheets, and embroidered clothes, a dresser, an electric fan, and dishes of daily use. Khudla's one-room house was the epicentre of happiness in the village that day.

Only a year later, a son was born to them, bringing the parents great joy. Another year later, a second son was born whom Khudla decided to give away to his childless brother to raise as his own, and Kalsoom did not oppose his decision. Soon, their son started going to the village madrassa (Islamic seminary) and the couple planned to build another room to extend their house. However, Kalsoom became pregnant with another child and gave birth to a daughter. Khudla never cared for his daughter like he cared for his son or his pigeons. Within a month of her birth, the child fell ill with serious stomach problems and had difficulty digesting her mother's milk. The town doctor suggested they buy powdered milk, a special baby formula that cost roughly 1400 Rupees ($9) a month. Khudla refused and ordered Kalsoom to feed the child cow milk instead. "He had all the money to feed his pigeons on almonds and pistachios but couldn't get baby formula (*ḍabe alā khīr*) for his daughter" Kalsoom complained, remembering the days when their daughter was ill with diarrhoea for months and Khudla only stayed with his pigeons.

In rural South Punjab, girls are still considered more of a "burden" (*bār*) on parents, as temporary residents who are destined to move to their husband's house in the patrilocal kinship system. They are also considered a "burden" because their marriage involves the provision of a dowry that sometimes involves absurd demands for cash, a new car, or even property by the groom's family. In addition, girls are a "burden" because their behaviour has to be closely regulated according to codes of honour and modesty. Any act that goes against local norms by a woman is likely to bring dishonour and shame to the father, the brother, and to other male relatives. It is considered the responsibility of men to regulate the mobility, behaviour, and practices of the women of the family, including daughters and sisters, and to kill them without a second thought if they bring dishonour to the family. Such cases of "honour killing" are not limited to South Punjab. Sharmeen Obaid-Chinoy's Oscar-winning documentary *A Girl in the River* (2015) and the works of many Pakistani and South Asian feminists suggest honour killing is practised all over rural Pakistan.

Khudla's selective neglect of his daughter was a part of his adherence to South Punjabi norms of hegemonic masculinity that had over time shaped his habitus to prefer his son over his daughter. However, Khudla's preference for his son also worried Kalsoom when the little boy started following in his father's footsteps, assisting Khudla every morning and afternoon in flying pigeons. When she complained about this, Khudla told her that the *shauq* of

Figure 5.1 A pigeon flyer's three-year-old son aspires to follow in his father's footsteps (photo by Mahar Akmal).

keeping pigeons is a productive activity that keeps children away from vices such as drinking, smoking, taking drugs, or stealing. This silenced Kalsoom since these morally denounced practices (drinking, stealing, or taking drugs) could potentially land the family in more difficulty.

In August 2017, Khudla became jobless when the village landlord he worked for decided to move to a nearby city to educate his children in an English-medium school. Initially, he tried to find some work but later, as Kalsoom noticed, "he started spending more and more time with his pigeons." She now realised the consuming nature of pigeon flying *shauq* that, like an addiction, was pushing her husband beyond performing his familial responsibilities. "Earlier, he used to fly pigeons only in the morning and in the afternoon but after losing his job he remained on the rooftop the entire day," she said. Spending money on feeding pigeons in times of hardship, she thought, was utter madness, "pigeons eat almost 300 Rupees daily [two kilograms of millet and 250 grams of almonds], and in return they give only *beṭh* (pigeon droppings)." The company of pigeons, which Khudla said brought him comfort in his hard times (see Chapter 2), Kalsoom considered was the main reason for his irresponsible behaviour. She expected Khudla to behave like a responsible man, to provide for the family and to find a job. The act of lavishing money on pigeons, she thought, was financially irresponsible and unmanly.

Over the past few decades, growing inflation, changes in the family structure, and minimal increase in the wages of labourers have greatly affected social life all over Pakistan. Recent price hikes in seeds, fertiliser, and pesticides have further exacerbated such uncertainty, while the price of crops such as cotton and wheat has not changed much in the market. For every menial job, there are a dozen labourers available. This means that although fifty years ago, one man was able to nourish a family of eight to ten individuals, today everyone in the house needs to work to maintain the domestic economy. Kalsoom did not have an extended family to care for (only a husband and two children), but she had to find work outside the house to supplement their income when her husband was unemployed. Along with raising children, cooking, cleaning, and performing other domestic chores, she started to pluck cotton in the fields, manually weed, harvest, and thresh crops, stitch clothes for money, and prepare and sell manure fuel plates (*thāpī*), clay hearths, and ovens. Khudla, meanwhile, remained on his rooftop all day long. When Khudla got two cows from a friend on a contract, he forced Kalsoom into taking care of them as well. Whenever she refused to do so, he would beat her. Such physical abuse, she said, did not make her husband a true man, "a man is responsible for the food, clothing, shelter and other necessities of his household members, whereas he is blind to his responsibility." Her definition of masculinity deviated from her husband's explanation of the concept who found true manliness in winning *izzat* through pigeon flying competitions. According to Kalsoom, "useless" (*faẓol*) activities like pigeon flying not only required "irresponsible spending" but also made the men idle because, instead of working to improve the domestic economy, they remained distracted by their *shauq*. When, in 2019, Kalsoom started keeping half-a-dozen hens in her house to sell their eggs for some extra cash, she became optimistic about her children's future and took on the role of provider and caregiver that are generally associated with men in rural Pakistan.

In South Punjab, men's aspiration to acquire hegemonic masculinity affects their familial and social relations. It is an ideal that men can adopt when it suits them and strategically discard when in other situations (Wetherell and Edley 1999). Khudla, for instance, demonstrated a different type of masculinity in the company of his pigeons. As he fed, bred, and decorated his birds, he indulged in effective gendered labour that made him care for his flock like a mother (see Chapter 2). However, away from the field of flying pigeon, in his everyday relations with family members, he consistently aspired to and demonstrated traits of hegemonic masculinity. As he involved his son in the activity of flying pigeons and decidedly ignored his wife and daughter, he tried to adhere to the accepted role of a man in his village. Since the structural ingestions of values of hegemonic masculinity organised according to the "androcentric principle" (Bourdieu 2001, 24) became an unconscious part of his habitus, it influenced his everyday decisions, familial relationships, and social practices.

Meanwhile, for Kalsoom, the "masculinity" of her husband was important but she wanted him to be a family "man" who preferred his family over his *shauq*, was responsible about his domestic duties, and was affectionate towards all children, daughters and sons alike. Some other wives of pigeon flyers had similar concerns, arguing that they were happy to support the household economy by working in the field or caring for the cattle, but they wanted their men to be "responsible" as a "protector" and a "provider." As flyers spent extended hours on their rooftops, ignoring the basic demands of their families and "squandered" time and resources on pigeons, Kalsoom and other women argued that this *shauq* eats up relationships like termites eat wood. The practice of *shauq* and care for the family, they believed, could be balanced, yet many men were unable to divide their time, care, resources, and affection between their non-human and human companions.

Kalsoom's life story is one among many such stories of women in rural South Punjab who show great determination in taking up family responsibilities and manage the household in the absence of a working man. For instance, when some men migrate to the Gulf states as labour, their wives ensure the education of children, care for elderly in-laws, manage the domestic economy, look after cattle, and maintain the well-being of all. Unfortunately, most women are still treated without respect and have to experience various forms of violence throughout their life. What separates Kalsoom's life story from others is her sharing of the domestic space with pigeons. While she observed Khudla's deep commitment to his pigeons and the selective neglect of her needs and that of their little daughter, she experienced the fragile, fluid, and contingent unfolding of life that is sometimes a characteristic of multi-species households (Charles 2014).

Finding honour in a multi-species household

When Sajida's marriage was arranged with her distant cousin, she was not aware of his craze for chickens. Akram, her husband, was a tall middle-aged man, the sixteenth and youngest child in the family who lived with his 75-year-old mother and a sister with mental health disability, after the death of his father. His six sisters and eight brothers were long married, but it was his obsession for chickens that meant no father in the village was willing to give him his daughter in marriage. When the hair on Akram's head turned more grey than black, his frail mother succeeded in arranging his marriage to a cousin fifteen years younger to him, a *pakī jaṭnī* (a strong village woman) who she thought would bring happiness to her son's life and take care of her and her daughter. Sajida was heavily built with flat feet and working hands, with a height of almost six feet. Since childhood, she had been trained to be a good housewife: to cook well, sweep the floor every morning, arrange charpoys, roll up the beds, dust the rooms, wash clothes and dishes, and cook again. For her, moving to an extended family after marriage or caring for an elderly mother-in-law and a sister-in-law was not an issue, "I used to care for my parents and younger sisters and brothers before marriage," she said. However, it was Akram's obsession with chickens that worried her. After a few days of marriage, she noticed how Akram's relatives sneered at him over his *shauq*. Her fears were confirmed when early one day, soon after their marriage, she cooked paratha and omelette for her husband's breakfast and he was furious with her after finding his *aṣīl* chicken's egg in the plate.

Sajida was lucky that Akram never insulted or physically abused her. He took great care of her when she was pregnant by making her *chorī* (kneading wheat bread by mixing sugar and *ghī*) and regularly calling the midwife for check-ups. When their first daughter was born, "he loved her like people love their sons," Sajida recalled. Two years later, another daughter was born to them and, after a miscarriage, two sons. However, his love for chickens was always the root cause of all troubles in the house. When Sajida finished cooking breakfast and cleaning the house, she needed to spend extra time on scrubbing the faeces-splattered brick floor. Often, just as she finished scrubbing, the "clever" chickens would lay their faecal load on the floor, and some would regularly ruin the charpoys and beddings with their droppings. What was worse, no one was ever allowed to eat their eggs or meat, and all eight chickens followed by their leader roaster, Nukrah, went about like they owned the house. Once when a friend asked Akram to give him Nukrah for some days to mate his hens and later informed him that he had lost the rooster in a cotton field, Akram did not eat for days. The thought of some jackal attacking and killing Nukrah deeply saddened him. Sajida remembered that he would leave the house early in the morning and spend all day in the cotton field, calling "O my Nukrah ... O my Nukrah." In those days, terrible silence (*chup*) engulfed the house and Sajida and her mother-in-law felt Akram's grief too. Later, as Akram's grief declined, he noted how many people in the village mocked his obsession and a village musician even composed a song called "O Nukrah ... O my Nukrah," mimicking Akram's sad tone when calling out to his beloved rooster.

For Sajida, this was a matter of family disgrace (*beztī*). Akram's act, she said, brought ridicule on the whole family, "Nukrah was only a rooster but Akram was sad like a person had died." She claimed that Akram's *shauq* usually impacted his self-esteem in the village, "when people refer to him as a *kukaṛ-bāz* (cockfighter) and a *jowārī* (gambler), it worries me." *Jowārī* is a derogatory term used for addicted gamblers in South Punjab, and when it is combined with *kukaṛ-bāzī*, it suggests an obsession or addiction that can lead to personal and financial ruin. Also, while Akram developed deep intimacy with his chickens and wagered by fighting this *halal* bird, Sajida believed the stigma extended to the entire household: "My two sons are identified as children of a cockfighter." She said this creates a bad identity (*burī pehchān*) for the family and make them appear less honourable (*ghaṭ izzat ale*) in the village. Sajida's argument is in line with the findings of Stephen Lyon who, in his ethnographic study of a Pakistani village, argues: "virtually everything a Punjabi individual does, has an impact on the family" (2004, 71), and that the honour of a man (or a family) is measured in part by what others say about him (2004, 64).

In rural South Punjab, it is considered men's primary responsibility to act and behave in a masculine manner to uphold family honour in the community. Wasting several days looking for a lost chicken or not eating in grief for not finding him, did not adhere to the values of hegemonic masculinity and made Akram, and the whole family in extension, look weak. According to Sajida, he intentionally ignored this reality and argued that *izzat* came from rearing good roosters and by winning cockfights. It was in 2018, when they arranged the marriage of their eldest daughter with Akram's sister's son that they encountered the harsh reality of what people thought of them.

In preparation for the wedding, Akram sold an acre of valuable land and spent the money on purchasing dowry and embroidered clothes for his daughter. However, before they could fix the date for the ceremony, the boy refused to proceed with the marriage because he said he wanted to marry respectable people (*waḍe lok*). This was a serious blow not only to Akram's *izzat* but also impacted Sajida's status among village women and their daughter's chances of getting a good proposal. Akram, who loved his daughter more than any other child, suffered a stroke and became bedridden. No one from the village came to their support. Sajida blamed Akram's *shauq* for bringing on this tragedy to their household:

> His *shauq* dishonoured (*beztī*) us in front of the whole village. What was wrong with my daughter? She is good looking, knows embroidery, and

would have served the *boa* (father's sister, the girl's future mother-in-law) all her life. The only problem is that she is the daughter of a cockfighter.

Akram was later diagnosed with a heart problem and hepatitis and, as his health declined rapidly, he quit cockfighting. Sajida thought it was too late in the day, "how did it matter now" (*hun kyā faīdā*) since they had already lost their *izzat* in the village. There are still chickens in the house but fewer in quantity since Akram is unable to look after them. He has asked Sajida to clean their coops and feed them twice a day, although this is a chore she resents.

Sajida's concern about the impact of a man's *shauq* on family's *izzat,* make us return to the question of embodied entanglements in an "inter-species social dwelling" (Campbell 2005, 79). More specifically, by following Charles (2014), I ask how affective relationships developed in the company of non-humans are experienced in a multi-species household. For Akram, chickens were important social actors who structured his daily routine, personal choices, and his conception of the self. As he fed and attended to his birds every morning and evening, this more-than-human relationship provided him profound joy and deep satisfaction. However, Sajida interpreted this differently from him. As an insider, she experienced the conflicting side of his *shauq* which meant dealing with his neglect of human relatives and witnessing his care of non-human companions. Akram's *shauq* marked her physical world too by making her rub the floors harder, clean the foul-smelling coops, and prepare feed for chickens. In her social universe, the birds were not passive objects but active subjects who caused a significant decline in the family's social capital. They were seen as a nuisance, as consumers of precious resources including time spent on looking after them, and as a reason for Akram's distraction from his social responsibility.

Worthy or unworthy *shauq*

This chapter discusses how human and non-human entanglements generated through *shauq* create domestic dilemmas in an inter-species social dwelling. As animal keepers prioritise scarce resources, time, and affection on their non-human companions over human family members, this results in women's strong objections to the men's *shauq*. Such objections are crucial in patriarchal South Punjabi society where many women regularly face structural and symbolic violence and have to rely on men's support, protection, and care. These women's lives are important and perhaps require a book-length project to critically investigate their agency, cultural negotiations, and gendered resistance. In this book, however, I emphasise a single aspect of their life to examine how they share their lifeworlds with the men's *shauq,* and how the presence of animals in or near a household affects their social universe.

The complex structure of care and neglect produced by the presence of animals within a domestic space leads us to ask whether pigeon flying, cockfighting, and dogfighting may be considered worthy *shauq*. According to the women, such pursuits perpetuate and even exacerbate household hardship, both materially and symbolically. Many women regard these practices as being wasteful (of time, money, and effort), or even more problematic, see them as an addiction (gambling), informed by immorality (a bad example for the children). They accuse their men of frivolous spending on animals in feeding and wagering on them, and label their *shauq* a wasteful activity. A dogfighter's mother argued that keeping a dog and feeding him is useless because he can neither be milked not eaten (*kutty nā ḍohwaṇ dy, nā kohwaṇ dy*). Her daughter corroborated, "I have begged my brother a hundred times to leave this useless *shauq* and keep cattle instead." The latter practice, she said, is a productive activity, with the prospect of a stable income and *izzat* in the community. Many women suggested that pigeon flying, cockfighting, and dogfighting were not only financially aberrant *shauq*, they also destroyed family honour in the community and affected personal relationships. For example, the wife of a dogfighter narrated how physically repelled she feels when her husband asks her to cook the meat as succulently for the dog as she would cook it for him. Such arguments, and the above three extracts from women's life stories, suggest that enduring entanglement between human and animal lives within a domestic space does not always result in peace and comfort, at times it generates disapproval from those who dislike living in the multi-species household.

For the *shauqeen*, their animals are cherished companions with whom they develop a diverse set of emotions over time. For instance, keeping pigeons, roosters, and dogs can generate joy and their illness or injury is a cause of grave concern. The masculine performance of these animals provides an opportunity to win honour among peers, and their skills help humiliate rivals. The men claim that an embodied connection with their animals makes them live a better life. That is why, when pigeon flyer Nadir's wife asked him to leave this *shauq*, he replied that he would accept anything else she had to say. "By looking at pigeons," he convinced her, "I forget about my sorrows" (Narayan and Kavesh 2019, 716). Such worries include social or family issues, or economic troubles, which Nadir felt he could ward-off when in the company of his cherished birds. Another pigeon flyer argued that involving his sons in taking care of the pigeons would inculcate stronger familial values in them. The rooftop was a space where he sought to bond with his sons and socialise them into what he thought were the etiquettes of life.[2] By flying pigeons, he stated, he was demonstrating to his sons the core values of caregiving (*khayāl*), hard work (*meḥnat*), and honesty (*imāndārī*). This practice also enabled him to teach his sons how to develop and maintain friendships, not succumb to bad acts (such as drugs, ogling women, stealing, and so forth), and to evaluate people on their merits and skills rather than on their social status. Similarly, cockfighters and dogfighters spoke about how participation in their activities strengthened core values in their sons, including an ability to courageously defend their honour in a challenging situation. In other words, animal keepers rebuffed the women's criticism by suggesting that dedication and passion towards a *shauq* (be it pigeon flying, cockfighting, or dogfighting) was an opportunity for personal growth and a way of nurturing the future generation.

When I visited Aslam one morning in autumn 2008, I observed a charpoy placed next to his pigeon coops. Squatting on the embankment of the canal that runs next to his adobe house, I asked him out of curiosity: "Do you sleep next to your pigeons?" He guffawed *"hān, ae vī mede bacy hen"* (yes, they [pigeons] are also my children). How Aslam equated his pigeons to his children was revealing; to him, caring for the birds was like caring for his children. Pigeons, like children, occupied his life in countless ways and he said he could not stop thinking about them, "walking, eating, meeting people, or sitting alone in my shop, I keep thinking about my pigeons." Later, he told me that he slept near his birds to protect them from thieves (who usually operate at night), and predators (such as cats and snakes) that kill pigeons. As a genuine *shauqeen*, it was his responsibility to ensure the safety of his birds. These threats were not always physical, sometimes they were encountered as the criticism of the villagers, the evil eye of an opponent, or the hazardous impacts of changing times. Addressing such threats was important for seasoned pigeon flyers, cockfighters, and dogfighters, as it ensured the continuity of their passionate interest that promised crucial symbolic rewards. I shall turn to address these threats to *shauq* in the final chapter.

Notes

1 Among the Sunni sect in Pakistan, according to prevailing interpretations of the Islamic law, uttering the word *talaq* three times nullifies the marriage.

2 While the focus of this study is largely on male relations (competitive or otherwise), other relations with women and children implicitly inform ideologies and perceptions of masculinities in rural Punjab. For instance, pigeon flyers stated that rooftops allowed them to educate their sons about life, values they believed in, and aspirations for the future. However, like many other South Punjabi men, pigeon flyers took very little interest in the socialisation of their daughters who were considered the wives' responsibility.

6   Threats to genuine *shauq*

In the introduction to the book, I suggested that Farid ud-Din Attar's *Manteq at-Tair* or *The Conference of the Birds* (1984) helps us better understand the social implications of *shauq* in rural South Punjab. The story that Attar narrates is about different species of birds who set out on a difficult journey in search of their king, the elusive Simorgh, after choosing the hoopoe as their leader. They wish to reach the court of their king but the journey is full of emotional and physical trials. Many of the birds end up forfeiting the quest, some surrender to the harshness of the way, while others are attracted by worldly desires. However, only thirty among them, the genuine *shauqeen*, show enthusiasm about reaching their objective and, at one stage, inquire of their guide about the length of the journey:

> "Before we reach our goal," the hoopoe said,
> "The journey's seven valleys lie ahead;
> How far this is the world has never learned,
> For no one who has gone there has returned"
>
> (Attar 1984, 166)

Shauq is their ultimate companion in this arduous journey which helps them push the limits to cross the seven valleys of: the quest (*talab*), love (*ishq*), knowledge (*ma'rifat*), detachment (*istighna*), unity (*tawhid*), bewilderment (*hayrat*), poverty, and annihilation (*faqr o fana*). The first valley is crucial since it helps them develop pure desire for the goal, and enables them to prepare for all uncertainties. At the start, the hoopoe advises them that reality may seem different and, as they enter the next stage, the valley of love (*ishq*), they experience a strong desire burning inside them that leads them to divorce from old habits and escape from the fears brought on by reasoning. This love then opens the doors to a different kind of knowledge (*ma'rifat*) which, unlike *'ilm* (also meaning knowledge) is not learned but experienced, witnessed, and unveiled, and can lead them towards the world of infinite possibilities. However, to reach their goal, the birds first detach (*istighna*) themselves from worldly pleasures and search for ultimate contentment in union (*tawhid*) with their king, Simorgh. This fifth stage is important as it allows the birds to go beyond their self-centred universe and actualise their *shauq* by surpassing their carnal self (*nafs*) and reasoning. Yet, this experience of union leads them to perplexity, a state of bewilderment (*hayrat*) that makes them question all previous logic, rationality, and associations. In this state of confusion, the thirty birds enter the last valley of poverty and annihilation (*faqr o fana*), and as their self re-emerges, they see themselves as Simorgh (*si* in Persian means thirty and *morgh* means birds). They come to know that their entire journey inspired by *shauq* was about the discovery of the self.

Attar's *The Conference of the Birds* is a mystical masterpiece that explains the various stages a wayfarer encounters in her quest for union with God (Stone 2006; Shackle 2006; Ritter 2003). The underlying motive that guides the birds' journey towards Simorgh or a mystic's search for Divine Love is *shauq*—dedicated passion that helps shape the practices of a person to experience or achieve something valuable. Such explanations of *shauq* were often given to me by pigeon flyers, cockfighters, and dogfighters, who rationalised their investment of time, energies, and money on their animals to achieve something for personal satisfaction. To this end, they also had to cross the seven stages to become "genuine *shauqeen*." For instance, when a pigeon flyer develops the desire (*talab*) to fly pigeons, he prepares himself by buying or building a pigeon coop on the rooftop and purchasing, breeding, and training pigeons. As he enters the next stage of love (*ishq*), he immerses himself in looking after his pigeons every morning and evening and goes beyond any fear. In this inter-species relationship, he develops an embodied knowledge (*ma'rifat*) through routinised intimate labour and care. However, as this *shauq* dominates the pigeon flyer's lifeworld, he starts experiencing detachment (*istighna*) from family members, neighbours, relatives, and other villagers, and gradually witnesses himself uniting (*tawhid*) with his pigeons. As he practises his *shauq* of keeping and flying pigeons amidst all familial and community criticism, he enters the state of bewilderment (*hayrat*), questions his choices and, most importantly, his intimacy with non-humans that results in emotional and physical distance from close family members. Very few men enter into the last stage where they annihilate (*fana*) their self in this *shauq* and experience a new self emerging, that of a genuine *shauqeen*. So, a part of my quest in this chapter is to explore this category of "genuine *shauqeen*" in the context of rural South Punjab.

"Only an honourable man possesses genuine *shauq*," dogfighter Ashiq told me in April 2015 as we sat together on a woven palm-leaf mat under a Neem tree in the open courtyard of his son-in-law's house, with golden wheat fields within sight. Almost 60 years of age, Ashiq had never been to school but I was not surprised by his use of the English word "genuine" to highlight the characteristics of "true" *shauq*. Many South Punjabis use the word "genuine" (with a slightly different pronunciation) in their daily conversations to hint at something authentic or pure, such as a "*genuine radio*" (original radio), a "*genuine banda*" (an authentic man), or even a "*genuine gāṇ*" (a pure-breed cow).[1] The term "genuine *shauq*" was one that Ashiq and many other animal keepers employed to express pure *shauq* practised to achieve personal grati-fication and joy even though it had no financial or material reward. Sitting cross-legged, Ashiq started explaining what seemed like a connection between honour and *shauq*. "A genuine *shauqeen* will always protect his animals from all threats," he added. When I asked about these threats, he explained three major categories: (1) threats from the villagers; (2) threats of the evil eye; and (3) threats of materiality and modernity. Ashiq believed that true enthusiasts should act like honourable (*izzat-dār*) men, devote their days and nights to their cherished animals, protect them from all possible harm, feed them well, and fly or fight them to admire their masculine abilities.

How does embodied association with animals generates threats? What is the nature of these threats? How do the animal keepers respond to such threats? And why is mitigating such threats crucial for the practise of "genuine *shauq*" and maintenance of one's masculine honour? My concern in this chapter is to examine the debates around threats to "genuine *shauq*." Following the explan-ation of Ashiq and other animal keepers, I divide these threats into three categories. First, I explore how different normative regimes try to suppress this community of enthusiasts. To unpack this, I adduce some ethnographic examples from my fieldwork and discuss how *shauqeen* men deal with the criti-cism of villagers and address other socioreligious pressures. Second, I explore how far a man may go to protect his animals and thereby save his *izzat* and masculine status. Since *shauq* is an intimate affair and because it involves an intricate relationship between a human and a non-human, the *shauqeen* often blame other people for casting the evil eye and bringing on misfortune to their animals. I am interested in the discourse and practice of safeguarding "genuine *shauq*" from the evil eye, and what recourse men take to address this. My final concern is with examining the wider relations forged through *shauq* in South Punjabi society where cultural values and ideals are in transition today. As mentioned in earlier chapters, *shauqeen* often speak of a degener-ating age where past ideals (such as respecting the wise *ustāds*) are continu-ously waning. My question here is what an enthusiast does to maintain the purity and sublimity of his "genuine *shauq*" from the effects of materiality and modernity.

Although the argument in this chapter revolves around *shauq*, its authen-ticity, vulnerability, and its evolution across time, at the heart of my concern is the human–animal relationships and their role in shaping men's concep-tion of honour (*izzat*) in rural South Punjab. I examine how genuine *shauq* renders pigeons, roosters, and dogs as active subjects of their human keepers' honour and masculinity, and how the progressive decline of *shauq* suggests that animals are becoming mere objects and docile bodies. This argument becomes clear in the last section of the chapter where I discuss how "genuine *shauq*" is affected by an enthusiast's decision to sell his cherished animal in the form of a commodity. As the animals are valued, kept close, and recognised with personalised names and traits, there is an expectation that they remain long-term companions rather than passive objects that are traded or sold in the market.

Fighting against norms

The general perception in rural South Punjab is that pigeon flying, cock-fighting, and dogfighting are useless activities (*fazol kam*). For the animal keepers, their *shauq* is a matter of holding on to a personal passion and zeal for the animal activities, to spend days and nights with their animals, and to fight them to win masculine *izzat*. Yet, many villagers speak of this passions with disdain and consider these as immoral practices that may lead a man to various normative violations and towards an abnegation of his social duties. In the early months of my fieldwork in 2014, as I climbed up to Rafik's rooftop to learn about the intricacies involved in pigeon flying, he started receiving complaints from his neighbours about my presence because an outsider male could easily violate *purdah* by looking into neighbourhood households and taking a peek at the women. Rafik had been living in this neighbourhood since decades and had a relationship of trust and respect with his neighbours, yet his guest was unwelcomed to such a space. A few months later, Ghafoor, a pigeon flyer in his early 50s who had lived his entire life in the residential blocks of Bukhari Chowk in central Kahror Pacca, received complaints from a neighbour for allegedly looking at his daughter while flying pigeons. A community meeting of elder men was summoned and Ghafoor was forced to either leave his *shauq* of pigeon flying or build a head-high wall over his rooftop for *purdah*-related reasons. Ghafoor agreed to build the wall. Both incidents led me to rethink the social status of pigeon flyers in the area, their vulnerability to such accusations and, most importantly, how this *shauq* combated old social norms such as *purdah*.

Purdah, literally meaning curtain, is a word commonly used by the South Asian Muslims to describe the system of secluding women to gendered spaces and limiting their mobility by enforcing values of modesty and shame (Papanek 1971, 517). In other words, the norm of *purdah* requires women to remain "inside" within the four walls of the house, and avoid public space or the "outside." The public space in South Asia, as Partha Chatterjee (1993) notes, is seen as the domain of the male, usually identified with materiality and danger, whereas the private space or "the home" is perceived as the centre of true identity and spirituality, that must remain unaffected by the "profane activities of the material world—and woman is its representation" (Chatterjee 1993, 120). The four walls of the house are considered a safe space for women, and the predatory gaze of outsiders at such spaces is regarded as a direct violation of *purdah* and as a violation of family honour. As many flyers like Ghafoor have to climb up to their rooftops to fly the pigeons, most rural South Punjabis interpret the flyers' act as their covert intention to peer into the neighbouring courtyards at unveiled women. Such considerations render pigeon flying a type of perversion, an un-Islamic and immoral activity that breaches household privacy and attacks the honour of women.

In response, Ghafoor and many other pigeon flyers argued that on the rooftop, they were completely focused on the flight of pigeons and had neither the time nor the intention of looking into other people's houses. They argued that their only love was pigeons who captivated their attention and made them unconscious of the outer world, as Rafik stated: "Pigeon flying is my love, and pigeons are my beloved. Day and night, I think only about them" (quoted in Narayan and Kavesh 2019, 711). He maintained that if this *shauq* is practised within limits and flyers spend sufficient time in seeking a livelihood in the day, then it is a productive rather than a destructive activity; "this *shauq* is not bad, but it gets a bad name (*badnām*)" [quoted in Narayan and Kavesh 2019, 722].

The norm of *purdah* is built on Islamic interpretations and its violation is considered a deviation from Muslim values. Another important normative value that is denounced by Islam is the practice of gambling for which many pigeon flyers and cockfighters are routinely castigated. Gambling (*jowā*), as I discussed in the previous chapter, is generally regarded as a corruptive and destructive practice that bears on a person's moral fibre. Like the violation of *purdah*, it is considered a sinful and amoral practice. Even worse, it is described as an addiction that reflects a man's inability to control his desires, leading him to waste money, and polluting various aspects of his social and family life. For such reasons, there is a strong legal restriction on gambling in Pakistan, which appeared in the last months of Zulfikar Ali Bhutto's reign in the late 1970s, particularly through the influence of "Nizam-e-Mustafa"

Figure 6.1 Pigeon flyers captivated by the sight of their airborne pigeons.

(Order of the Prophet)—an Islamist movement supported by right-wing religiopolitical parties (Richter 1979, 548).

In describing the obsessive and destructive nature of gambling, many villagers narrated tales of affluent pigeon flyers and cockfighters who ruined themselves by purchasing expensive birds and gambling on them. While narrating such stories, people drew a cautionary conclusion about overindulging in *shauq* that could prove to be destructive of one's economic and social life. However, many pigeon flyers and cockfighters disagreed with this criticism and argued that gambling was not for winning money; it was an expression of commitment to their *shauq*. Such arguments are in line with Rebecca Cassidy's (2010, 140) study of gambling houses in Britain where she suggests that the question is not how much money a person wins or loses but "what does it mean to win or lose?" Winning or losing money through gambling over pigeons and cocks was far less important to the *shauqeen* since the main purpose of making a wager was to build networks, demonstrate expertise to peers, and fulfil *shauq*.

The denunciation of cockfighters was compounded by what the villagers described as cruelty (*zulm*) and sinful (*gunāh alā*) acts where roosters were primed to fight for human enjoyment. Many people believed that cockfighters were cruel to speechless creatures (*be-zubān makhloq*) who could not complain about their tormentors. Here again, religious dictums were summoned in support of the argument and animal fighters were seen as violating the teachings of the Prophet who strictly forbade fighting, baiting, or harming animals in any way.[2] Speaking about cockfighting, Moen, a cloth merchant in his 30s from the city of Lodhran, argued, "Allah has great dislike for people who harm other creatures just for fun, and will blind them [cockfighters] like their birds [who sometimes are blinded in combat]." He recalled a story about a blacksmith, Hanif, a famous cockfighter in the last two decades of the twentieth century. As he got older, he lost his eyesight and everyone knew it was due to his past deeds.

The concern with animal cruelty was far less of an issue with dogfighters. Here criticism revolved around the care dogfighters bestowed on a "polluted" and "impure" animal. The dog, as discussed in Chapter 4, is a *haram* animal in Islam, usually associated with dirt, filth, and pollution, and treated indifferently in most parts of South Punjab. Because dogfighters touched this "unholy" animal, fed him dairy products such as milk, butter, and *ghī*, and kept him within or near their house, they were often criticised for such proximity. From the day I started my fieldwork, I was cautioned by several villagers to refrain from meeting or sharing food with dogfighters whose reputation as dirty (*gande*) people preceded them. Speaking about dogfighters, a person from the village Khan da Kho said, "It is a matter of extreme disgust and revulsion when dogfighters carry their *haram* animal on their shoulders after the fight." He labelled this *shauq* as a curse, "they are cursed (*phiṭ he inhānko*) to keep dogs, otherwise, no sane person would feed pure products (milk, butter and *ghī*) to a dog." In the Qur'ān, milk has been described as "palatable to drinkers" (16, 66), and the faithful are promised "rivers of milk" in paradise (47, 15). In Pakistan, milk (and other dairy products) is considered pure and sacred (Philippon 2012, 295; and Mughal 2014, 103; Parkes 1987, 653), thus leading many people to believe that the provision of this sacred substance to "polluted" dogs is abhorrent and disrespectful.

All these criticisms are important and develop a particular discourse among the village people regarding pigeon flying, cockfighting, and dog-fighting. Such discourse shapes people's perceptions and ideologies regarding the category of *shauq* and its capacity to cultivate *izzat*. As the father of a dogfighter argued,

> A *shauq* is to feed a bull, make him bulky and strong, so when he is sold he earns a stack of money (*thabbī raqam dī*) and brings *izzat* in the village, whereas keeping a dog only brings loss (*charā nuqṣān*).

Caring for livestock, as discussed in the previous chapter, is considered far more productive, both socially and economically, than keeping a dog which may be seen as a wasteful activity. The concept of waste in South Asia, as Vinay Gidwani (1992) notes, carries the meaning of what is socially undesirable: as a noun, it is marginal, unimportant, or unpalatable, and as a verb, it is associated with human behaviour that is unacceptable and unproductive (1992, 31). In this sense, both as a noun and as a verb, the *shauq* (or "wasteful activity") of fighting dogs is viewed as the primary reason for the social and economic downfall of the *shauqeen*. For instance, Rana Abdul-Rehman, a famous dogfighter, sold almost eleven acres of valuable agricultural land to purchase champion dogs. For him, many village people argued, dogfighting was not only a wasteful activity but also a curse, a threat to his moral and socioeconomic well-being.

The relentless attitude of the animal keepers makes many village people argue against calling these activities "*shauq*." Instead, they describe these activities as an addiction because they are based on an uncontrollable obsession that makes men forget their responsibilities towards their children and family. Many villagers used pejorative terms to refer to animal keepers like *nashā'ī* (addicts), and drew a comparison between them and chronic hashish consumers (*charsī*). Some villagers also argued that the *shauq* of flying and fighting animals was like a disease (*bemārī*) that was impossible to cure, and that travelled across generations and contaminated the larger community.

The list of such criticism is long; however, I have listed some major themes to explain how "genuine *shauqeen*" respond to such normative threats. The men regarded their *shauq* as a serious pursuit rather than a pleasurable pastime and argued that their animals were important actors in their lives whose presence brought them personal satisfaction and social prestige or *izzat*. Cockfighters reiterated that betting was not for the sake of earning money but to stimulate the pleasure of fighting cocks—an act that people without *shauq* could not understand. Dogfighters, too, gave a number of reasons for

Figure 6.2 Four generations—Manna (second from right), accompanied by his father (far left), son (second left), and grandson (far right). Both Manna and his son were dogfighting enthusiasts, while his father usually criticised their *shauq*.

why they kept "polluted" dogs. Ashiq, for instance, argued that the dog is also "a creature of Allah (*Allah di makhloq*), has life (*sāh*), in winter he feels cold, in summer he feels hot, and like humans, he also requires food to live on." He did not consider himself a "cruel" (*ẓālim*) man but a "dog lover" who kept and fed dogs despite strong social disapproval. Dogfighters argued that dogs possess three vital traits that are fast disappearing in human society: loyalty (*wafā*), obedience (*adab*), and honesty (*imān-dārī*). Therefore, Basheer argued, he was proud to keep a dog, because "there is more sincerity in the dogs than in humans these days." Riaz summarised: "*to kutte ko hik wārī roṭī ghat, oh wal na bhoṇksī* (once you feed a dog, he will never bark [at you])." Such observations suggest a deeper understanding of why this *shauq* is attractive to people in the first instance.

Although most dogfighters described the loyalty of dogs as the prime reason for keeping canines, however, merely keeping dogs was not their *shauq*. The *shauq* was to fight the animal and showcase his stamina, dominance, valour, and aggression in the presence of a crowd of thousands. Many saw their *shauq* as an act to gain *izzat* by winning the competition and by humiliating their village rivals. Their *shauq*, in other words, was not about building an intimate connection with a *halal* or *haram* animal but fighting an animal that represented the values and ideals of dominant masculinity.

Yet, almost all pigeon flyers, cockfighters, and dogfighters accepted that spending time with their animals was an escape from the moral decadence of present times and a way to achieve solace (*sakon*), which they said was rarely available in the company of humans. They justified their decision to keep animals in the house by arguing that their non-human companions did not discriminate between rich or poor, Sunni or Shia, Punjabi or Seraiki, and local or Mohajir, they just loved humans with devotion and sincerity. They argued that the loyalty and unconditional love found in animals is missing in humans and added that, while a person may be a trickster (*chāl-bāz*), an animal's intentions are always straightforward. Many of the men reported that while close relatives and even children left them in hard times when they suffered a loss in fortunes, their animals remained loyal to them. The category of *shauq* continues to resonate strongly in the lives of the animal keepers in an environment that is increasingly disapproving of their activities. Some men told me that after the massive Islamisation of the country in the 80s, their *shauq* gradually became less acceptable to people around them, yet they held on to their passion because it brings meaning to their lives. They suggested that it is not simple to give up their activities because they are victims of an all-embracing *shauq*: "*shauq dy hatho majbor hen*" (we are helpless to our *shauq*).

Above, I have discussed how animal keepers try to maintain their *shauq* despite constant criticism from the rural community. However, one disconcerting aspect of *shauq* that the men raised was the fear of the evil eye that others could cast on them unintentionally and bring misfortune and ill health to their cherished animals. Many animal keepers also believed that the intimate bonds between the *shauqeen,* such as being friends (*yār*) and brothers (*bhirā*), could be damaged by envy and jealousy channelled through the evil eye. Below, I examine how the evil eye is considered a threat to animals, to animal keepers' social relationships, and to *shauq* and what the men do to ward off such harm.

Saving *shauq* from the malicious gaze

Because their animals are vehicles of their *shauq*, the men prioritise their well-being and protection from all physical and cosmological threats. As discussed in previous chapters, before flying or fighting competitions are organised, the men spend hours in carefully preparing the diet of animals and in training them for combat. However, if despite all careful preparation, the animals do not perform well or die, one common explanation is that the animal has fallen victim to *nazar*, or the evil eye. The belief is that the evil eye is unintentional, and the inflictor is unaware of the harm she is causing. Yet, it is believed to lie in the innate and uncontrollable gaze of jealous people.

Belief in the evil eye, as folklorist Alan Dundes (1981) explains, spreads across many cultures of the world. In Muslim societies, this belief is stronger because it finds support in textual Islam.[3] In South Asia and among the South Asian diaspora, as many ethnographies show (e.g., Pocock 1973, 25–40; Shaw 2000, 200–201; Callan 2007, 332–333; Spiro 2005, 61–68; Rytter 2010, 47; Woodburne 1981, 55–65), this belief shapes everyday relationships and practices. For example, David Pocock (1973) in his ethnography of the Patidar people of central Gujarat in India, found that belief in the evil eye served to reinforce caste and social hierarchies. What functional explanation, we may then ask, does belief in the evil eye provide to the rural South Punjabi animal keepers, and how does it influence their relationship with their animals, with peers, and affect their *shauq*.

According to many animal keepers, the envious gaze or praise of others can work maliciously and worsen the condition of humans, animals, crops, and objects such as houses and cars. Many misfortunes are attributed to the evil eye; for instance, it can make children sick, ruin a newly married couple's life, destroy houses and crops, cause significant loss to a prosperous business, or death to an animal, an accident to a new car, and so forth. South Punjabi belief in the evil eye can be categorised under two major types: malignant gaze (*ghairān dī nazar*), and emotional praise (*apnyān dī nazar*). The first type is the jealousy and envy of others (*ghair*) about someone's prosperous condition, while the second type is the affection and appreciation of close ones (*apny*). These are two sides of the same coin, as both can result in harm, even if there is no explicit malicious intention. In explaining the second type of evil eye, animal keepers give the example of a mother who can harm her child with too much praise, the overt manifestation of deep love among spouses can lead to a separation, and the appreciation of one's animals can affect their health and performance. However, as discussed above, the perpetrator is often difficult to identify, and even if someone is accused of casting the evil gaze, it is assumed that the person is unaware of the damage she is inflicting.

Interestingly, even though many pigeon flyers believed in the evil eye, they argued that the pigeon never fell prey to it because the bird was a "holy" animal who preferred to dwell in shrines and in the minarets of mosques; and was therefore shielded from cosmological harm. In addition, some flyers believed the pigeon was a "Sayyid bird"[4] who has witnessed the martyrdom of the nephew of the Prophet, Imam Hussain, at Karbala where it dipped its beak in the blood of Imam Hussain and flew back to Medina to inform the people about the tragedy (see also Frembgen and Rollier 2014, 51). Because of such strong religious symbolism attributed to the bird, most of my interlocutors believed that the evil eye could not affect the pigeon. However, good breed roosters and dogs were susceptible to the evil eye, as the case of Waso's rooster illustrates.

Do not praise him: Nazar on Waso's rooster

"I tried every cure but Hitler's health got worse and worse, and he died within three days." Waso's tone carried an undercurrent of sorrow as he narrated the story of his most cherished gamecock, Hitler. Although the rooster died in 2010 (before my fieldwork with cockfighters began), his image was vivid in Waso's memory. Removing a loose-fitting turban from his head, Waso recounted, "Hitler exhibited the qualities of a champion as soon he hatched from the egg." He was the quickest and strongest among all the chicks, and when Waso tried him in sparring exercises with senior cocks, "his jumps were high, his attacks were sharp, and his balance was extraordinary." Considering Hitler's remarkable qualities, Waso spent countless hours in training and feeding him. When Hitler was eighteen months of age, he took part in a cock-fight and killed his adversary in only three minutes. At this time, someone from the crowd shouted, "the rooster is cruel like Hitler" and this name became the cock's identity. Waso nostalgically recounted, "When Hitler used to kick the opponent in fury, the opponent died instantly. His kick was so powerful." Hitler's fame soon spread, and everyone was praising the cock's skills. This was the time when Hitler fell prey to the evil effects of praise.

All good gamecocks are susceptible to the evil eye, as was the case with Hitler. To protect him from the infective gaze and blight of jealousy, Waso often carried him under his arm, buried in a black shawl, away from the sight of others. The cock was only released publicly in the pit. One day, Waso's cousin who was also a cockfighter and coveted Hitler, came visiting unexpectedly and showered endless praise on the gamecock: *"ae kukar barā changā he"* (this rooster is so good). Waso was unable to deflect the evil gaze of his cousin. Later that night, Hitler fell sick. Waso knew that Hitler had fallen victim to the evil eye but he needed to confirm this. He asked his wife to put some red chillies in a tray and swirl it over Hitler's head before throwing the chillies into the fire. When there was no emanation from the fire, his worse fears were confirmed. To counteract this, he did "everything": he kneaded seven balls of wheat-flour and swirled them over Hitler's head in a clockwise direction and threw them over the wall; he tied black threads around the cock's shanks; he soaked an old leather shoe in water and give it to Hitler to drink. When all such measures failed and he found no improvement in the bird's condition, Waso attempted what he called *"dwā ty d'uā"* (biomedical treatment and prayer). He took Hitler to a veterinary doctor and started biomedical treatment. He also took Hitler to the local Imam of the mosque and asked him to pray for his health. Waso did not sleep for three nights, doing everything in his power to save Hitler, but the cock only grew worse and died. Waso said he had many gamecocks, but none of them was quite like Hitler.

The perpetrator of the evil eye is hardly ever accused; however, it still indicates the jealousy of the inflictor towards one's happiness and prosperous condition. Waso blamed his cousin's envy for Hitler's death, who had always coveted his rooster. In other words, as Waso projected his anxiety about possessing a champion cock, he found the evil eye to justify his misfortune. This afflictive supernatural power, he believed, was a major threat to his *shauq*, and could potentially have an injurious effect on his cherished rooster. Therefore, when Hitler got sick, Waso tried all folk remedies for the evil eye but did not consult the vet until the day the cock died. This was also the point of many veterinary doctors I interviewed who argued that no matter how often they urge the villagers to seek instant medical consultation for sick animals, they waste time by treating animals for the evil eye. This, they said, only worsens the sick animal's condition.

Hitler's case study is not a rarity, the belief in the evil eye is enmeshed in the everyday social relations of almost all cockfighters and dogfighters. Even though strong bonds such as friendship (*yārī*) and brotherhood (*bhirapī*) are held in esteem, cases like this can lead to hatred among the men. Moreover, such beliefs help the men explain the inexplicable, including unexpected defeats or the death of an animal. By referring to the evil eye as the cause of misfortune, the men attempt to avoid public humiliation. In this way, belief in this cosmological force helps them preserve their masculine *izzat* by justifying their loss, and thus protects the integrity of their genuine *shauq*.

However, a question emerges why, unlike pigeon flyers, cockfighters and dogfighters are partial to supernatural explanations for their losses. There are two possible explanations. First, pigeon flocks generally comprise around 100–150 pigeons and the loss of one or two pigeons does not affect the flyer's *shauq* as heavily. Whereas roosters and canines are kept in much fewer numbers and any loss greatly affects the continuity of a man's *shauq* and his standing among peers. Second, a normal gamecock or fighting dog is expensive to buy (20,000–80,000 Rupees on average), whereas pigeons come cheap (200 Rupees). Since cocks and dogs involve a significant outlay of funds by the keepers, their loss warrants an explanation.

Overall, the evil eye is seen as a serious threat to genuine *shauq*, which is believed to bring misfortune on cherished animals and affects a man's relationship with peers. However, *shauq* faces another grave challenge. In the context of changing South Punjabi society, the *shauq* of flying and fighting animals is on the wane. Seasoned animal keepers often speak of such changes in terms of authentic or "genuine *shauq*" that has been compromised by newcomers who breed animals for monetary gain—a topic I turn to next.

Transformations

"There are two types of pigeon keepers," Ustad Nadir told me while artfully maintaining the balance of his old motorbike as we drove slowly through the half-grown wheat fields. "Some are genuine *shauqeen*, and others are *chiṛī mār* (literally, sparrow hunters, or greedy captors, as Nadir conceptualised them)," he continued.[5] It was a pleasant February afternoon in 2015 and Ustad Nadir was taking me to watch a *khokhā* pigeon competition. Sitting on the rear seat of the roaring two-stroke Yamaha, I tried hard to catch his words. "Pigeon flying was a passion of the Mughal kings, right?" he asked and I nodded. "After the long day of politics in court," he continued, "the Mughals used to relax by looking at and flying colourful pigeons." Their beautifully patterned pigeons were a "window of solace" (*sakoon dī khīṛkī*), who helped them forget their worries. However, Ustad Nadir suggested, whenever a king's pigeon landed on the rooftop of a commoner, the king would send his royal courier to retrieve the bird and give some money to the poor person in recompense. According to him, since the reward offered was a considerable sum, impoverished people began catching and keeping pigeons as a way of making a living. Ustad Nadir equated the past with the present by calling "genuine *shauqeen*" to be like the Mughal emperors who kept colourful pigeons and found peace and joy in them, while "greedy captors" were like the poor people who practised catching pigeons as a *kārobār,* a lucrative business. To a "genuine *shauqeen*," Ustad Nadir argued, every bird had a distinct personality; some were cherished for their valuable colours and patterns, while others were loved for their loyal character. However, he and some other old-timer pigeon flyers suggested, there was now a significant increase in the numbers of greedy captors who had made it their business to catch other peoples' pigeons. Because of these "inauthentic" (*ḍo-number*) pigeon flyers, they suggested, the *shauq* is losing its charm.

The *shauq* of finding something significant for yourself, as Attar tells us through the precarious journey of his birds, requires utmost commitment and purity of desire. At the start of *The Conference of the Birds,* Attar introduces many birds who wish to seek their king, however, the nightingale is in love with the world, the rose, and the garden; the parrot is arrogant and fears death; the partridge pursues material wealth; the vanity of the peacock makes him selfish; the falcon follows the false king; the owl is happy inhabiting ruins, and so forth. Only thirty birds possess pure desire and under the leadership of hoopoe set on the journey to find their king Simorgh. These thirty birds have a strong commitment and are therefore resistant to all temptations of material wealth, artificial pleasures, and fake desires. Such commitment and purity of desire turns the birds into genuine *shauqeen* and they, like a seasoned South Punjabi enthusiast, cross all seven stages to discover their true self.

Many South Asian scholars have argued that a major trait of *shauq* is that it must be practised with purity of desire and without regard for material wealth.[6] It is a personal predilection "that draws people to particular cultural practices not because they *have* to but because they *want* to" (Narayan 2016, 224 my emphases). Since the purpose is personal gratification, pursuits inspired by *shauq* represent a pleasant space of action beyond pecuniary interests (Narayan and Kavesh 2019). In rural South Punjab, I also found that many animal keepers perceived *shauq* in opposition to the labour performed for financial gain. However, this zest for flying or fighting animals is changing in modern times with the dominant discourse marked by the corrosive effects of the unbridled desire for money. Many *shauqeen* I spoke to suggested that if an activity is practised for pecuniary benefit, it ceases to be *shauq* and becomes *kārobār*. *Kārobār* can be glossed as "business," or as any task that is performed for profit. The seasoned animal keepers (or old-timers–who have been keeping animals for generations) use this juxtaposition of *shauq* and *kārobār* to castigate "unserious" animal keepers. According to them, such men have lost their way and indulge in "disgraceful acts" of making money by selling their animals.

For pigeon flyers, as Ustad Nadir's comments illustrate, the pursuit of money signifies that a person is not a "genuine *shauqeen*" but an "inauthentic" greedy captor. These greedy captors include those *khokhā* (coop) pigeon flyers who sell their entire lot (almost 100 pigeons in one coop) in exchange for money or property. According to the old-timers, such transactions mean the men never developed a close attachment to their birds and do not appreciate their individual characteristics. The category of greedy captors also includes flyers who exploit the bird's fecundity for laying eggs every month, and breed pigeons on a large scale to sell them in the market for the money. However, some pigeon breeders avoid buying from the market and prefer person-to-person exchange.

Fateh Baig is one such pigeon trader and is referred to as *wapārī*. The term *wapārī* can be translated as a "merchant" but, among pigeon flyers, it has a derogatory meaning—a greedy person whose economic transactions are not trustworthy. Fateh Baig proudly informed me that he went to remote villages to buy novel coloured pigeons which he then sold to rich, urban pigeon flyers on a good profit. However, no matter how much money Fateh Baig made in his business, other flyers were unimpressed by this and considered him a greedy and inauthentic *shauqeen.* The old-timers blamed such greedy captors for dishonouring the values of "genuine *shauq.*" Ustad Jameel noted that in the past, "... people used to spend all day on rooftops in the heat of July to witness and praise the flight of pigeons." However, he argued, such traditions were now ebbing away. Similarly, Rafik sighed, "The genuine *shauqeen* are very few now, and indecent people (*lumb*) have taken over this activity." He believed that some money-oriented flyers had tarnished the reputation of the whole community. "These men are greedy captors who violate *purdah,*" he remarked. Such acts made people condemn all pigeon flyers and regard this *shauq* as a frivolous activity.

Like pigeon flying, the *shauq* of cockfighting is also undergoing transformation. Although it involves betting, as detailed in Chapter 3, these wagers serve a particular goal: to boost the thrill of the fight and to demonstrate the support and confidence of friends and supporters. Since the bets are small, no more than $6 for a single fight per person, the money is seen as a means rather than an end in itself. Winning such a bet does not make a person rich and losing it does not render him poor. Cockfighters are adamant that such regulations mean that despite the presence of money and betting, the genuine passion associated with the *shauq* of cockfighting is not compromised. However, there are people who violate this code and are criticised by other cockfighters for their general disregard of roosters, their personalities, fighting styles, and winning histories. Like pigeon flyers, some cockfighters claimed "genuine *shauq*" and regarded themselves as seasoned enthusiasts who continued to uphold the codes of honour. To them, their *shauq* is invaluable and requires commitment, devotion, and serious hard work. To take pleasure in the courageous combat and acrobatic stunts of fighting cocks, they argue, men must invest weeks and months to prepare their birds for the fight. After such investment, the high point of their *shauq* is achieved when the cock outwits his opponent through his fighting skills and receives applause from the cockfighting audience. However, if this pursuit for joy and pleasure (*shauq*) is contaminated by the pursuit of economic capital (money) through betting, the commitment and authenticity of the person's *shauq* comes under question. The older cockfighters argued that "unserious" newcomers, those who were not accustomed to the traditions of this *shauq* and just adopted it for fun, tends to ruin the purity of cockfighting *shauq*.

Dogfighters also believed that pecuniary interests ruin the sublimity (*shān*) of *shauq*. When Basheer Laar sold his two-year old dog, Bruce Li, for 80,000 Rupees (about $500), many dogfighters were shocked and expressed their surprise at Basheer's "disgraceful" decision. The disappointment was compounded by the fact that Basheer was regarded as a true enthusiast. His decision to sell Bruce Li not only affected his *izzat* and status among dogfighters but also raised questions even among close friends who began doubting his "genuine *shauq*." When I visited Basheer, he told me that he decided to sell his dog because a good amount of money was offered to him. He conceded it was a tough decision to make but argued, "In present times, it is useless to spend endless amounts on a dog without getting any profit from him." Basheer's views mirror a changing landscape wherein famous dogfighters are willing to breach age-old practices and norms. Whether this type of profit-making can be reconciled with *shauq* is hard to answer, however, in present times, many old-timers believe pecuniary interest is a major threat to *shauq*.

Older animal keepers consider mixing money with *shauq* as morally decadent for multiple reasons. Firstly, money downplays the value of hard work (*mehnat*). It is precisely because of a man's efforts in breeding, feeding, and training the animals that even poor animal keepers can compete against the rich. However, if prized animals can be purchased easily, it undermines the value of hard work and allows the rich to possess the best coloured pigeons, skilful cocks, and powerful dogs, with little investment beyond money. Similarly, by purchasing factory-produced animal feed, rich men surpass the effort of those enthusiasts who spend hours preparing the animal feed. Such standardisation and commodification is a mark of capitalism, displacing traditional social relations and cultural practices. The objective of making money from the animals, many old-timers believe, ruins true passion because it impedes a man from developing a bond of intimacy with his animals over years of close association.

Secondly, the business of buying and selling the animals makes the merchant richer, but it renders the seller morally vacuous. Although those who buy the animals cheaply from the impoverished animal keepers get a good deal, the man who sells his animals to the merchant experiences moral loss and develops a feeling of disengagement and detachment for his cherished animals. In this case, money depersonalises long-term relationships developed between humans and non-humans. It also generates regret and hatred and threatens friendly bonds between enthusiasts.

Lastly, the exchange of pigeons, roosters, and dogs commodifies the animals and renders them mere objects, whereas for a "genuine *shauqeen*" they are active actors and inalienable companions who promise honour and glory. Through his *shauq*, a man develops deep knowledge of his animals (such as during breeding and training), tends to their needs (through feeding), and experiences delight (through flying or fighting). However, when the animals become a type of commodity, this turns the birds and canines into objects that are unable to evoke strong feelings in their human caretakers.

Most animal keepers agreed that a market-driven approach would ruin "genuine *shauq*." *Shauq* was a productive endeavour that extended beyond the realm of the animals where new social relations were formed and sustained, wise masters were respected, friends were cherished, and opponents were humiliated. However, now this "genuine *shauq*" was threatened by the march of modernity and the insatiable desire for money. This desire was particularly associated with young animal keepers who viewed non-humans as commodities that could be traded rather than cherished. Many old-timers argued that young animal keepers lacked the innate inclination that is crucial to developing *shauq*. By hinting at his inborn attraction for pigeons, Rafik criticised the young flyers, "I was almost two-years-old when I first saw pigeons in my uncle's house. I ran after them madly, even though I had barely learnt how to walk. Now, young boys do not have that inclination (*wo cheez*)." By criticising young people, Rafik and other seasoned flyers tried to re-establish their connection with an idealised history and practice of pigeon flying.

External factors such as inflation, consumerism, demographic and urban expansion, modern education, and the role of the media were all factors listed as reasons for the decline of "genuine *shauq*." As I outlined earlier, older dogfighters like Basheer had started converting their *shauq* into a livelihood. While discussing about Basheer's decision to sell his cherished dog, Ashiq expressed empathy: "It is difficult to blame Basheer. You can see the increase in the cost of milk, butter, and *ghī*. How can you feed a dog?" He added that enthusiasts need to work extra hours to support their *shauq*, and predicted, "... soon people will have no time or money for dogfighting." In addition to the influence of money, the value of time has changed drastically. Urban expansion in recent decades has given birth to a labour class from rural areas whose work is measured in capitalist standards of time and money (see Thompson 1967). Moreover, the educated middle-class is scornful of such activities and considers these a sign of Seraiki backwardness and illiteracy. It is not surprising that those who still nurture *shauq* are aware of its imminent end. Old-timers predict that, in a couple of decades, these animal activities will completely disappear, along with the values they uphold. No longer will anyone remember the passion to train an animal for combat, the excitement of the competition, the expression of honour, the portrayal of masculinity, and the shame and humiliation of defeat.

Exploring genuine *shauq*

In his famous work, *Argonauts of the Western Pacific* (1932), Bronislaw Malinowski suggested, "we have to study man, and we must study what concerns him most intimately" (1932, 25). He further argued that in each culture, people have different aspirations, aims, and modes of happiness. Before studying cultural institutions, anthropologists must take into account "the subjective desire of feeling by what these people live," and "the substance of their happiness" (1932, 25). *Shauq* in the context of South Punjab is that subjective feeling and substance of happiness through which animal keepers live and shape their social worlds, achieve honour, and reproduce traits of hegemonic masculinity. It is a philosophy of life through which they structure their daily routines, develop friendships and intimacies with their non-human companions, and defend their passion from the corrupting effects of the evil eye, as well as that of modernity and market exchange. It is also what allows them to cross seven valleys like Attar's thirty birds and discover inner delights despite everyday economic troubles.

In this chapter, I have discussed the threats to "genuine *shauq*" in three ways. First, I described social threats to *shauq* conveyed by villagers including the neighbours and relatives of the animal keepers. Second, by a short case study of Waso's rooster, I examined how belief in the evil eye leads to enmity and hatred among men and influences their social relationships. Third, I explicated the tensions between *shauq* and *kārobār* (business) and detailed how animal activities are undergoing structural transformation. In all three sections, I highlighted some crucial characteristics of "genuine *shauq*" and explained how this can help us better understand human–animal relationships in the South Punjabi cultural context.

Throughout my fieldwork, the study of *shauq* helped me in building a relationship of trust and mutual respect with my interlocutors. When Malinowski advised the ethnographer to sometimes put aside the camera, notebook, and pencil and to "take part in the natives' games," and "sit down and listen and share in their conversation" (1932, 21), he did not only mean how to develop good rapport with the natives, but also how to develop a relationship of intimacy, friendship, and mutual attachment with the locals based on participant observation. Malinowski practised this and remarked, "I have carried away a distinct feeling that their behaviour, their manner of being, in all sorts of tribal transactions, became more transparent and easily understandable than it had been before" (1932, 22).

When I started taking interest in the *shauq* of my interlocutors, they began to open up, shared their stories, and introduced me into the intricacies, specialities, and technicalities of their enthusiasm and their subculture. The animal keeper's *shauq* became a frame through which I was able to portray their social landscapes. As I engaged with them on a daily basis, sometimes watching them peacefully adorning their pigeons and at other times fighting cherished roosters and dogs, my ethnographic study of their lives became my own passion, enthusiasm, and *shauq*.

Notes

1 In his introduction to "Keywords," anthropologist Craig Jeffrey argues that many English words are used casually in South Asia to convey different meanings at different times and in a different context (Jeffrey 2017, 272–273).
2 See Fakhar-i-Abbas (2009, 55–56) and Frembgen and Rollier (2014, 144).
3 In the Qur'ān, the evil eye has been mentioned (68, 51; 113, 5), and many Hadiths (Prophet's sayings) discuss its harms.
4 The term Sayyid is generally used for descendants of the Prophet.
5 In pigeon flying language, *chiṛī mār* are greedy pigeon flyers who do not think pigeons as distinct beings. They take the activity not as a passion, but only to capture other people's pigeons, a demeaning act much like hunting sparrows.
6 For some works that emphasise this distinction between *shauq* and money, see Baily (1988, 101), Sakata (1983, 84), Marsden (2005, 130), Wilkinson-Weber (1999, 132–135).

Epilogue
Life beyond cage and leash

In June 2017, I revisited my field site for a short, one-month trip and met friends and animal keepers. Early one evening, I found Rafik surrounded by customers at his famous kebab stall in Kahror Pacca. Clad in a thin white sleeveless singlet and grey trousers, he was absorbed in cooking the round *shami* kebabs. His absolute devotion to his work reminded me of our first meeting when he was just as deeply immersed in flying his flock of pigeons. After some time, as he glanced over the customers and met my gaze, Rafik asked his son to take over the stall and moved to greet me warmly. We relaxed on two worn chairs under the fluorescent light beside his kebab cart and chatted about life, work, and family. A noisy pedestal fan threw air and provided us some relief from the heat. Rafik showed me his "pigeon note-book": a detailed record of the breeding, caring, training, feeding, and flying techniques of the birds. The book also contained recipes for healing potions, and passages on the appropriate distribution of various seeds according to the seasons. "This is the hard work of my entire life, the sum of my mastery (*ustādi*)," Rafik remarked as he held on to the hand-bound note-book. "And yet," he acknowledged soberly, "there is no one I can pass on this knowledge to." Rafik's despair was palpable, frustrated at the lack of interest of young pigeon flyers who, he claimed, did not show any eagerness to learn or master the skill of flying pigeons. Leaning forward, he said, "my two sons have the *shauq* of pigeon flying too, but it is different from my generation. They take this *shauq* like a sport, not as a matter of their *izzat*." He questioned the devotion of young flyers; their apparent indifference to the pigeons' needs, and overall reluctance to invest the hard work and time required to sustain this *shauq*, including cleaning pigeon coops, respecting old masters, and observing the *purdah* norm. "Times are changing," Rafik mused, "pigeon flying is now losing its meaning."

Next morning, as the news of my arrival spread in the village, dogfighter Makhan Pehlwan invited me to his father-in-law's adobe house. Other friends and relatives of Makhan joined us and we lounged on four charpoys, placed in opposite directions to each other. The government had not yet supplied power to that area, so most people used palm leaf hand-fans to cope with the oppressive heat of the afternoon. To my embarrassment, Makhan's younger brother, Abid, a muscular wrestler in his 20s, was ordered by his brother to stand and swing a hand-fan for my comfort. My embarrassment grew as I was presented with halva made of *desī ghī*. This was a sign of great respect, indicating that my interest in the *shauq* of dogfighting was appreciated by the men. They recognised that, unlike a city news reporter, I did not see dogfighting as a frivolous pastime of rural Seraikis, but rather as something that deserved attention and careful study.

After discussing life in Australia, our discussion turned to the subject of dogfighting. Makhan and his father, Ashiq, told me that in their area dogfights had decreased in recent times. "As you know," Makhan said, "the government is becoming increasingly strict and putting an end to all public gatherings." He was referring to the chain of terrorist attacks on holy sites in recent years that provoked authorities into banning festivals and other public gatherings held around rural shrines. Makhan added that he ventured out to faraway places in search of a dogfight but could not find one. Makhan and other *shauqeen* continued to fight their dogs to defeat village rivals and accumulate culturally important symbolic rewards. However, he argued, "as this *shauq* is now declining, there is no *izzat* in keeping dogs."

Later that week, I met with cockfighter Imran Niazi who cheerfully bought me a glass of sugarcane-juice from a local vendor. As we sat near the city's canal gulping the fresh energy drink, Imran spoke of the sadness that gripped him after the death of his cockfighting master. As a devout disciple, he attended the funeral prayers but now that the wise *ustād* was no longer there to give his blessings and advice on matters related to the cockfight, Imran lost much of his passion and stopped keeping roosters. He spent most of his time and energy on his work, trying to earn extra money to build a house, prepare dowry for his young daughter's marriage, and send his sons to an English-medium school. He acknowledged that he still harboured the *shauq* and sometimes attended cockfighting but, with the disappearance of old cockfighters, the activity had become more about acquiring money through betting and less of an opportunity to achieve *izzat* among peers. "The *shauq* is there," he said with a broad smile, "but cockfighting is changing."

There seemed a consensus among South Punjabi animal keepers that cultural meanings associated with the *shauq* of pigeon flying, cockfighting, and dogfighting were changing. As I visited the animal keepers again in April 2018, and later contacted them over the phone, I noticed their frustration with the lack of seriousness among young people for these "old" (*purāny*) pursuits, and new restrictions by the state. Almost all of my interlocutors, the ones who devoted their energies, time, and money to their animals, lamented these changes, casting themselves as the last generation of *shauqeen*. To these men, their *shauq* adorned their lives and injected a thrill into the mundane, daily struggle of fulfilling social and familial obligations. It helped them forge meaningful relationships with their animals and with fellow enthusiasts, altered their experience of life and work, and gave them a sense of personal well-being. Above all, their *shauq* facilitated in outmanoeuvring experienced animal keepers and humiliating rivals with the objective of achieving *izzat*.

How does the idea of *izzat* shape people's everyday life choices, practices, and their knowledge of self and non-human others in rural Pakistan? I opened this book with a short story of Raza, a young pigeon flyer who forced me to rethink this question and look at the South Punjabi concept of *izzat* from a fresh perspective. His observations inspired me to explore the role of *izzat* in shaping the life choices of rural South Punjabi men, and understand its critical difference from common dictionary translations such as "honour, prestige, pride, reputation." Although I have largely translated *izzat* as "honour" in this book for the ease of a general reader, I argue that this cultural concept generates "relational complexity" (Haraway 2016, 20) between humans and non-humans, structures the *shauqeen's* social values, and forms their symbolic expectations. In this way, *izzat* "becomes almost tangible: it can be gained or lost, preserved or squandered" through its everyday retelling and renegotiation (Liechty 2003, 84). Whenever the men competed their pigeons, roosters, and dogs in the arena, I noticed their *shauq* made *izzat* appear as tangible as their animals.

The book explored ethnographically how we may understand the South Punjabi concept of *izzat* through the *shauq* of keeping, flying, and fighting animals. *Shauq*, I contended, animates the animal keepers' lives, structures their priorities, enhances their sociality, and allows them to develop a meaningful relationship with non-humans. In illustrating diverse modalities of inter-species intimacy through the breeding, feeding, and training of animals, my point is not that such *shauq* is innocent, but that this cultural concept enables us to explore complex entanglements of care and violence, affection and neglect, love and indifference that knot humans and animals lives in rural Pakistan. To put it differently, my attempt to understand the accumulation of *izzat* through the cultivation of *shauq* may not explore the lifeworlds of animals but it helps us understand more-than-human entanglements that, in the words of Donna Haraway (2016, 29), provide "a possible thread in a pattern for ongoing, noninnocent, interrogative, multispecies getting on together."

Pigeons, roosters, and dogs have histories of becoming-with human beings in South Asia. Chapter 1, "Decolonising Passions," explored such histories and discussed how, for many centuries, these animals have tied people into knots of place, gender, race, class, and other social phenomena. Pigeons, for instance, have animated the lives of Indian emperors, kings, and rulers for centuries whereas, in colonial times, roosters and dogs were staged to fight to help the British and native Indian elites build important political relationships. Through their flight or fight, these animals enabled men to display masculine pride, reiterate their symbolic dominance, and achieve important political objectives. In the latter half of the nineteenth century, the practice of flying pigeons or fighting roosters or canines began to lose its status among the ruling elites, largely due to the influence of various political and social developments taking place in Victorian England, along with the increased prevalence of Anglo-Protestant masculine traits, the rise of the animal welfare movement, and the enforcement of the *Prevention of Cruelty to Animals Act* in India (1890).

In present-day Pakistan, these animals appear to have a complex relationship with the post-colonial state. Until 2018, there was no revision to the colonial version of the *Prevention of Cruelty to Animals Act* (1890), and the 2018 amendment only changed the amount of penalty levied on a lawbreaker, without altering a single world of the *Act*. It still permits people to "incite animals to fight if ... all reasonable pre-cautions are taken to prevent injury" (Section 6-C). Since the suffering of animals is not accounted for and their intrinsic value is not recognised, this *Act* only serves as a relic of colonialism in contemporary Pakistan. If there is a recent decline in dogfighting, as Makhan's comments suggest, it is primarily to ban public gatherings and not dogfights. This leads us to question the state's attitude which, instead of developing clear legislation for animal welfare, leaves activities like cockfighting and dogfighting as a distinctive "trait" of rural Pakistan. Roosters continue to fight and dogs are still baited, and police raids of animal fights in South Punjab only reinforce the idea of Seraiki backwardness in popular discourse. In this way, echoing Akhil Gupta (2012, 77), the state regenerates the categories of South Punjabi "poor" and Seraiki "backwardness."

The relationship between humans and animals that emerge through specific historical and political processes, now shape the lives of many rural South Punjabi men. As the *shauqeen* breed pigeons on their rooftops, carefully prepare feed for their roosters, or train their canines, their everyday attachment with animals allows them to fulfil their enthusiasm and achieve something meaningful. Chapter 2, "Living with Pigeons," unpacked this claim of multispecies relatedness by discussing how the men indulge in intimate labour and clean the pigeon coops, carefully decorate the birds' feet with distinct beads and bells, provide them with an environment of ideal mating conditions, appreciate their unique colours and patterns, and train them to turn into loyal companions. Such provision of care to non-humans is also evident among cockfighters and, as Chapter 3 "The Seduction of Cockfighting" explored, it enables the men to consider their roosters as beings with distinct personalities and fighting styles. In their *shauq*, many cockfighters spend a great deal of time with their gamecocks, breed them carefully, raise them responsibly, give them unique names, keenly prepare their feed, and train them to fight with determination, courage, bravery, and strength. The care of canines, however, is different since it involves social and moral sacrifices for the men to develop a close relationship with a ritually polluted, unclean, and *haram* animal. "The Spectacle of Dogfighting" (Chapter 4) explained how the men, despite strong religious and cultural restrictions, cross symbolic boundaries to feed, train, and massage their fighting dogs. Such relationships with pigeons, cocks, and dogs, I have argued, facilitate the men in attaining joy in the company of their non-human companions and achieve something "good" by momentarily forgetting the vagaries of everyday life.

The aim of "an anthropology of the good," Joel Robbins (2013, 457) convinces us, "is to explore the different ways people organise their personal and collective lives in order to foster what they think of as good." The good, he argues, is something that is "imaginatively conceived" and to study it, anthropologists

> need to be attentive to the way people orientate to and act in a world that outstrips the one most concretely present to them, and to avoid dismissing their ideals as unimportant or, worse, as bad-faith alibis for the worlds they actually create.
>
> (2013, 457)

I have taken such an approach to study animal intimacies developed in rural South Punjab and discarded my personal preferences and ethical positions to prioritise explanations that my respondents provided me about keeping, breeding, and competing animals. In this project of an anthropology of the good, I learned from my interlocutors that the choice to fly or fight animals is not a form of cruelty to animals but *shauq*, not a sport but a matter of *izzat*, and not a poor man's hobby but the well-established passion of kings.

However, this turn to anthropologies of the good, as Sherry Ortner (2016, 60) argues, is tangled with "dark anthropology," or the one that focuses on power, suffering, inequality, pain, and violence. In paying closer attention to the manner in which more-than-human entanglements generate different knots of relatedness, I also experienced that "relatedness always entails some kind of violence" (Govindrajan 2018, 178). The *shauq* of keeping pigeons, roosters, and dogs was not limited to rooftops or households but extended beyond the "domestic realm" and allowed the men to indulge in a struggle to accumulate honour and social prestige through the masculine performance of their animals. Although pigeon flyers and cockfighters also strived for masculine honour, among dogfighters the struggle for such symbolic gains was pronounced. This was in part because of the large sums of money invested in fighting dogs, and because dogfights were held as public events in front of thousands of spectators who monitored and judged the courage and strength of the competing canines and their keepers. As the animals fought, bled, and died in front of my eyes in their quest to recapture their keeper's honour, I witnessed the existence of violence in multi-species relatedness.

The ethnographic cases of pigeon flying, cockfighting, and dogfighting I have discussed in this book can be taken as a lens to look at larger cultural values, including how *shauq* helps the men perform different masculinities in the village setting. Masculinity performed through pigeon flying, for instance, is a non-violent one; it does not require "bravery" or physical strength, and yet, within their field, a skilled pigeon-flyer is seen as a man of honour. Unlike pigeon flying, the masculinity associated with cockfighting places far more emphasis on attributes such as courage, strength, and bravery and in the pit, both the fighting cock and his keeper are expected to display these traits. Dogfighting encourages a hegemonic form of masculinity. Betting and any other pecuniary rewards are prohibited in this activity as dogfighters bring their canines into the fighting arena to display and appreciate the animal's aggression, domination, and valour.

All these masculinities are performed and valued as ideal manly practices to achieve *izzat* among peers in the changing social and physical landscape. However, the wives of the animal keepers castigated such practices and argued that that role of an "ideal" man is to protect, care for, and support his family. Chapter 5, "A Life with *Shauqeen*," explored how the women had an alternate interpretation of ideal manliness as they coped with the experience of neglect, indifference, and violence while witnessing their husband's care, affection, and attention to their non-human companions in a multi-species household. In such cases, I argued that the notion of masculine honour in the "domestic" realm—as breadwinners, protectors, and care givers—was in tension with its meaning in the "public" realm, where the men sought to assert their masculinity through their *shauq*.

It is perhaps because these pursuits are carried out with great passion and commitment that they face serious threats in the modern age. "Threats to Genuine *Shauq*" (Chapter 6) discussed why the *shauq* is becoming increasingly hard to sustain in rural South Punjab and how seasoned enthusiasts or old-timers try to maintain a commitment to their pursuits. In general, old-timers criticised the obsession of young, money-oriented animal keepers who breed and sell animals like a commodity for pecuniary benefit. They call the younger generation *burger lok* (burger persons), decrying their love of fast food (the iconic American burger) over traditional ways of eating, and *mobile lok* (mobile persons), referencing the obsession that young people have for mobile phones and virtual interaction (see Doron and Jeffrey 2013). Such materialism, they maintain, is not simply a sign of individual failing (greed) but a wider social malaise afflicting Pakistani society. They use local references to describe this shift; for instance, in a world where *desī ghī* (pure clarified butter) was once valuable, it has been replaced with hydrogenated oil; where brown domestic chicken's eggs were prized, white battery farm eggs are everywhere; where real wood has been replaced with chipboard; and *shaker* (raw brown sugar) with white sugar. While attributing all these changes to the corrosive effects of materialism and modernity, many old-timers argue that as modernity eats into social relations and contaminates cherished practices, it will also affect people's ways of living in the world and their conceptualisation of honour.

What can we learn from the tension between newer and older generations in rural South Punjab regarding cultural practices like animal keeping? While this may be viewed as inter-generational friction often found in societies experiencing rapid social change that poses a threat to structures of authority and established social networks, such debates are also reflective of wider transformations around cultural categories of *shauq* and *izzat*. Rafik, Makhan, and Imran's views at the start of this epilogue reiterate this point and suggest that, as *shauq* transforms among people, it also affects their markers of *izzat*. The opinions of these men emphasise the fluidity of the concept of *izzat* which, as Steve Taylor (2013, 400) suggests, changes with time and space, and displaces older aspirations with new motivations. As the concept of *izzat* changes, it will accommodate new forms of engagements and alter the conception of self in society. Such changes in *izzat* and *shauq* are already visible in rural South Punjab as rich landlords replace their finely bred horses with new sedans and SUVs, and bear-baiting dogs with greyhounds as markers of their *izzat*; village cattle herders try to gain prestige by keeping an outcross of Jersey cows rather than *desī* cattle; and rural youth prefer playing cricket and volleyball over flying pigeons. Whatever the *shauq* may be, it will continue to enliven the social experiences of people in culturally meaningful ways and shape their shared understanding of *izzat* in society.

I end my discussion by acknowledging that while exploring different engagements between humans and animals, the important question that I have left unexplored and one which is worthy of a separate research project is the lifeworlds of animals. Pigeons are usually forced to fly during the unbearable summer heat and chickens and dogs often face life-threatening injuries. These critters are unable to express their suffering, even as the exhausted bodies of pigeons, limping cocks, and fighting scars on dogs' faces betray a harsh reality. I do not mean to underplay the painful life trajectories of the animals. However, as mentioned in the introduction, my key concern revolved around following the animal keepers' enthusiasm and wider understanding of what constitutes *shauq*, *izzat*, and *mardāneat* (masculinity) in rural South Punjab. By discussing the men's experience of these concepts, I have tried to uncover an understudied cultural phenomenon and argued that men's passionate attachment to their animals in rural South Punjab, albeit developed in ethically complex ways, informs their understanding of the self, allows them to withstand hardships, develop friendships and rivalries, and contest masculine honour. With such an objective, this book asks us to critically evaluate how animals, when they go beyond the cage and the leash, enliven life in rural Pakistan.

References

Abu-Dawud. n.d. "Sunan Abu Dawood." Hadith of Prophet Muhammad. http://sunnah.com/riyadussaliheen/18/220.

Affergan, Francis. 1994. "Zooanthropology of the Cockfight in Martinique." In *The Cockfight: A Casebook*, edited by Alan Dundes, 191–207. Madison: University of Wisconsin Press.

Ahmed, Akbar S. 1980. *Pakhtun Economy and Society*. London: Routledge and Kegan Paul.

———. 2002. *Discovering Islam: Making Sense of Muslim History and Society*. New York: Routledge.

Al-Bukhari, Muhammad. n.d. "Sahih Al-Bukhari." Hadith of Prophet Muhammad. http://sunnah.com/bukhari/59/111.

Al-Dameeri, Allama Muhammad Kamal-ud-Din. 2006. *Hayat-ul-Haiwan (Life of Animals)*. Translated in Urdu by Mulana Abdur Rasheed. Vol. 1. Lahore: Maktabh-ul-Hassan.

Ali, Ahmed. 1966. *Twilight in Delhi*. London and New York: Oxford University Press.

Alter, Joseph S. 1992. *The Wrestler's Body: Identity and Ideology in North India*. Berkeley: University of California Press.

———. 2002. "Pehlwani: Indian Wrestling and Somatic Nationalism." In *Combat, Ritual, and Performance: Anthropology of the Martial Arts*, edited by David E. Jones, 81–98. Westport, Connecticut, London: Praeger.

Anderson, Nancy. 2010. *The Sporting Life: Victorian Sports and Games. Victorian Life and Times.* Santa Barbara, California: Praeger.

Appadurai, Arjun. 1996. *Modernity at Large: Cultural Dimensions of Globalization*. Minneapolis and London: University of Minnesota Press.

Attar, Farid ud-Din. 1984. *The Conference of the Birds*. Translated by Afkham Darbandi and Dick Davis. London: Penguin Books.

Avieli, Nir. 2012. "Dog Meat Politics in a Vietnamese Town." *Ethnology: An International Journal of Cultural and Social Anthropology* 50 (1): 59–78.

Bailey, P. 2007. *Leisure and Class in Victorian England: Rational Recreation and the Contest for Control, 1830–1885*. London and New York: Routledge.

Baily, John. 1988. *Music of Afghanistan: Professional Musicians in the City of Herat*. Cambridge: Cambridge University Press.

Banerjee, Sikata. 2005. *Make Me A Man!: Masculinity, Hinduism, and Nationalism in India*. New York: State University of New York Press.

Barkat, Samiullah. 2009. *Kalam Hazrat Baba Bulleh Shah*. Okara: Karmawala Book Shop.

Barth, Fredrik. 1959. *Political Leadership among Swat Pathans*. London: The Athlone Press.

BBC. 2015. "Pakistani 'spy Pigeon' Arrested in India." Newsbeat. London: BBC. 2015. www.bbc.co.uk/newsbeat/article/32928909/pakistani-spy-pigeon-arrested-in-india.

Beebe, William. 1921. *A Monograph of the Pheasants*. Vol. 2. 4 vols. London: Witherby & Co.

Benthall, Jonathan. 2007. "Animal Liberation and Rights." *Anthropology Today* 23 (2): 1–3.

Besnier, Niko, and Susan Brownell. 2012. "Sports, Modernity, and the Body." *Annual Review of Anthropology* 41: 443–59.

Best, Steven. 2009. "Manifesto for Radical Abolitionism: Total Liberation by Any Means Necessary." Animalliberationfront.Com. 2009. www.animalliberationfront.com/ALFront/Manifesto-TotalLib.htm.

Bio-resource Research Centre (BRC). 2010. "Bearbaiting." Islamabad: Bio-Resource Research Centre (BRC). 2010. www.pbrc.edu.pk/bearbaiting.htm.

Bird, Sharon R. 1996. "Welcome to the Men's Club: Homosociality and the Maintenance of Hegemonic Masculinity." *Gender & Society* 10 (2): 120–32.

Bourdieu, Pierre. 1977. *Outline of a Theory of Practice. Cambridge Studies in Social Anthropology 16*. Cambridge and New York: Cambridge University Press.

———. 1984. *Distinction: A Social Critique of the Judgement of Taste*. Cambridge, Massachusetts: Harvard University Press.

———. 1986. "The Forms of Capital." In *Handbook of Theory and Research for the Sociology and Education*, edited by J. G. Richardson, 241–58. New York: Greenwood.

———. 1990. *The Logic of Practice*. Cambridge, UK: Polity Press.

———. 2001. *Masculine Domination*. Stanford and California: Stanford University Press.

Brohi, Nazish, and Afiya S. Zia. 2012. "With the Will to Die: Agentive Defiance to Honour Codes in Pakistan." In *Honour and Women's Rights: South Asian Perspectives*, edited by Manish Gupte, Ramesh Awasthi, and Shraddha Chickerur, 115–40. Maharashtra: Masum.

Brownstein, Oscar. 1969. "The Popularity of Baiting in England Before 1600: A Study in Social and Theatrical History." *Educational Theatre Journal* 21 (3): 237–50.

Bulliet, Richard W. 2005. *Hunters, Herders, and Hamburgers: The Past and Future of Human-Animal Relationships*. New York: Columbia University Press.

Callan, Alyson. 2007. "'What Else Do We Bengalis Do?' Sorcery, Overseas Migration, and the New Inequalities in Sylhet, Bangladesh." *Journal of the Royal Anthropological Institute* 13 (2): 331–43.

Campbell, Ben. 2005. "On 'Loving Your Water Buffalo More Than Your Own Mother': Relationships of Animal and Human Care in Nepal." In *Animals in Person*, edited by John Knight, 79–100. Oxford and New York: Berg.

Candea, Matei. 2010. "'I Fell in Love with Carlos the Meerkat': Engagement and Detachment in Human–Animal Relations." *American Ethnologist* 37 (2): 241–258.

Cassidy, Rebecca. 2002. *The Sport of Kings: Kinship, Class, and Thoroughbred Breeding in Newmarket*. Cambridge and New York: Cambridge University Press.

———. 2010. "Gambling as Exchange: Horserace Betting in London." *International Gambling Studies* 10 (2): 139–49.

Chakrabarty, Dipesh. 1992. "Of Garbage, Modernity and the Citizen's Gaze." *Economic and Political Weekly* 27 (10/11): 541–547.

———. 1999. "Adda, Calcutta: Dwelling in Modernity." *Public Culture* 11 (1): 109–145.

Charles, Nickie. 2014. "'Animals Just Love You as You Are': Experiencing Kinship Across the Species Barrier." *Sociology* 48 (4): 715–730.

Chatterjee, Partha. 1993. *The Nation and Its Fragments: Colonial and Postcolonial Histories.* Princeton: Princeton University Press.

Cheney, Margaret. 1981. *Tesla: Man out of Time.* New York: Marnes & Noble Books.

Chopra, Radhika, Caroline Osella, and Filippo Osella. 2004. *South Asian Masculinities: Context of Change, Sites of Continuity.* New Delhi: Women Unlimited: An Associate of Kali for Women.

Coleman, Phyllis G. 2009. "Note to Athletes, NFL, and NBA: Dog Fighting Is a Crime, Not a Sport." *Journal of Animal Law & Ethics* 3 (1): 85–119.

Connell, Raewyn W. 1995. *Masculinities.* Berkeley: University of California Press.

Connell, Robert W, and James W Messerschmidt. 2005. "Hegemonic Masculinity Rethinking the Concept." *Gender & Society* 19 (6): 829–59.

Crapanzano, Vincent. 1986. "Hermes' Dilemma: The Masking of Subversion in Ethnographic Description." In *Writing Culture: The Poetics and Politics of Ethnography*, edited by James Clifford and George E. Marcus, 51–76. Berkeley: University of California Press.

Csikszentmihalyi, Mihaly. 2014. *Flow and the Foundations of Positive Psychology: The Collected Works of Mihaly Csikszentmihalyi.* Claremont: Springer.

Dalrymple, William. 2002. *White Mughals: Love and Betrayal in the Eighteenth-Century India.* New York: Viking.

Darwin, Charles. 2010. *The Variation of Animals and Plants Under Domestication.* Cambridge: Cambridge University Press.

Dickens, Charles. 1851. "Spitalfields." *Household Words* 3: 25–30.

Donlon, Jon Griffin. 1996. "Cockfighting." In *Encyclopedia of World Sport: From Ancient Times to the Present*, edited by David Levinson and Karen Christensen, 1:186–91. Santa Barbara: ABC-CLIO, Inc.

Doron, Assa, and Robin Jeffrey. 2013. *The Great Indian Phone Book: How the Cheap Cell Phone Changes Business, Politics, and Daily Life.* Massachusetts: Harvard University Press.

Douglas, Mary. 1966. *Purity and Danger: An Analysis of the Concepts of Pollution and Taboo.* London: Routledge and Kegan Paul.

Dundes, Alan. 1981. *The Evil Eye: A Casebook.* Madison: University of Wisconsin Press.

———. 1994. *The Cockfight: A Casebook.* Madison: University of Wisconsin Press.

Evans-Pritchard, E. E. 1940. *The Nuer, A Description of the Modes of Livelihood and Political Institutions of a Nilotic People.* Oxford: Clarendon press.

Evans, Rhonda D, and Craig J Forsyth. 1997. "Entertainment to Outrage: A Social Historical View of Dogfighting." *International Review of Modern Sociology* 27 (2): 59–71.

———. 1998. "The Social Milieu of Dogmen and Dogfights." *Deviant Behavior* 19 (1): 51–71.

Evans, Rhonda, DeAnn K Gauthier, and Craig J Forsyth. 1998. "Dogfighting: Symbolic Expression and Validation of Masculinity." *Sex Roles* 39 (11–12): 825–38.

Ewing, Katherine. 1984. "Malangs of the Punjab: Intoxication or Adab as the Path to God?" In *Moral Conduct and Authority, The Place of Adab in South Asian Islam*, edited by B. D. Metcalf, 357–71. Berkeley: University of California Press.

———. 2010. "Anthropology and the Pakistani National Imaginary." In *Beyond Crisis: Re-Evaluating Pakistan*, edited by Naveeda Ahmed Khan, 531–40. New Delhi: Routledge.

Fakhar-i-Abbas. 2009. *Animal's Rights in Islam*. Saarbrücken: VDM Verlag Dr. Muller Aktiengesellschaft & Co. KG.

Fazl, Abul 'Allami. 1873. *Ain-i-Akbari*. Translated by M. A. H. Blochmann. Calcutta: Baptist Mission Press.

Feeley-Harnik, Gillian. 2007. "'An Experiment on a Gigantic Scale': Darwin and the Domestication of Pigeons." In *Where the Wild Things Are Now*, edited by Rebecca Cassidy and Molly Mullin, 147–82. Oxford and New York: Berg.

Fijn, Natasha. 2011. *Living with Herds: Human-Animal Coexistence in Mongolia*. New York: Cambridge University Press.

Fischer, Michael D. 1991. "Marriage and Power: Tradition and Transition in an Urban Punjabi Community." In *Economy and Culture in Pakistan: Migrants and Cities in a Muslim Society*, edited by Hastings Donnan and Pnina Werbner, 97–123. New York: Palgrave Macmillan.

———. 2006. "The Ideation and Instantiation of Arranging Marriage Within an Urban Community in Pakistan, 1982–2000." *Contemporary South Asia* 15 (3): 325–39.

Foucault, Michel. 1988. "Technologies of the Self." In *Technologies of the Self: A Seminar with Michel Foucault*, edited by Luther H. Martin, Huck Gutman, and Patrick H. Hutton, 16–49. London: Tavistock Publications.

Francione, Gary L. 2004. *Rain Without Thunder: The Ideology of the Animal Rights Movement*. Philadelphia: Temple University Press.

———. 2010. "The Abolition of Animal Exploitation." In *The Animal Rights Debate: Abolition or Regulation?*, edited by Gary L. Francione and Robert Garner, 1–102. New York: Columbia University Press.

Francione, Gary Lawrence, and Robert Garner. 2010. *The Animal Rights Debate: Abolition or Regulation?* New York: Columbia University Press.

Fraser, David. 2009. "Assessing Animal Welfare: Different Philosophies, Different Scientific Approaches." *Zoo Biology* 28 (6): 507–18.

Frembgen, Jurgen Wasim, and Paul Rollier. 2014. *Wrestlers, Pigeon Fanciers, and Kite Flyers: Traditional Sports and Pastimes in Lahore*. Karachi: Oxford University Press.

Frembgen, Jürgen Wasim. 2006. "Honour, Shame, and Bodily Mutilation, Cutting Off the Nose Among Tribal Societies in Pakistan." *Journal of the Royal Asiatic Society of Great Britain & Ireland* 16 (03): 243–60.

———. 2008. "Marginality, Sexuality and the Body: Professional Masseurs in Urban Muslim Punjab." *The Asia Pacific Journal of Anthropology* 9 (1): 1–28. https://doi.org/10.1080/14442210701822183.

———. 2011. *At the Shrine of the Red Sufi: Five Days and Nights on Pilgrimage in Pakistan*. Karachi: Oxford University Press.

Galaty, John. G. 1989. "Cattle and Cognition: Aspects of Maasai Practical Reasoning." In *The Walking Larder: Patterns of Domestication, Pastoralism, and Predation*, edited by Juliet Clutton-Brock, 215–30. London: Unwin Hyman.

Geertz, Clifford. 1972. "Deep Play: Notes on the Balinese Cockfight." *Daedalus* 101 (1): 1–37. https://doi.org/10.2307/20024056.

———. 1973. *The Interpretation of Cultures*. New York: Basic Books.

Gibson, Hanna. 2005. "Dog Fighting: Detailed Discussion." Michigan Staff University, College of Law: Animal Legal & Historical Center.

Gidwani, Vinay Krishin. 1992. "'Waste' and the Permanent Settlement in Bengal." *Economic and Political Weekly* 27 (4): PE39–46.

Government of Pakistan. 1977. "The Prevention of Gambling Act." XXVIII. Pakistan: Government of Pakistan.

———. 2018. *Amendment the Prevention of Cruelty to Animals Act, 1890.* www.na.gov. pk/uploads/documents/1526547955_562.pdf.

Government of the Punjab. 2015a. "Forestry, Wildlife and Fisheries." Lahore: Punjab Code. 2015. www.punjabcode.punjab.gov.pk/index/listdept/d/Forestry,+Wildlife+a nd+Fisheries.

———. 2015b. "Livestock and Dairy Development." Lahore: Punjab Code. 2015. www.punjabcode.punjab.gov.pk/index/listdept/d/Livestock+and+Dairy+Develop ment.

Govindrajan, Radhika. 2018. *Animal Intimacies: Interspecies Relatedness in India's Central Himalayas.* Chicago and London: University of Chicago Press.

Guggenheim, Scott. 1994. "Cock or Bull: Cockfighting, Social Structure, and Political Commentary in the Philippines." In *The Cockfight: A Casebook*, edited by Alan Dundes, 133–73. Madison: University of Wisconsin Press.

Gupta, Akhil. 2012. *Red Tape: Bureaucracy, Structural Violence, and Poverty in India.* Durham and London: Duke University Press.

Haraway, Donna Jeanne. 2003. *The Companion Species Manifesto: Dogs, People, and Significant Otherness.* Paradigm. Chicago: Prickly Paradigm Press.

———. 2008. *When Species Meet.* Minneapolis and London: University of Minnesota Press.

———. 2016. *Staying with the Trouble: Making Kin in the Chthulucene.* Durham and London: Duke University Press.

Harris, Jim. 1994. "The Rules of Cockfighting." In *The Cockfight: A Casebook*, edited by Alan Dundes, 9–16. Madison: University of Wisconsin Press.

Herzfeld, Michael. 1985. *The Poetics of Manhood: Contest and Identity in a Cretan Mountain Village.* New Jersey: Princeton University Press.

———. 2004. *The Body Impolitic: Artisans and Artifice in the Global Hierarchy of Value.* Chicago: University of Chicago Press.

Herzog, Hal. 2010. *Some We Love, Some We Hate, Some We Eat: Why It's So Hard to Think Straight About Animals.* New York: Harper.

Hirsch, Dafna. 2015. "Hygiene, Dirt and the Shaping of a New Man Among the Early Zionist Halutzim." *European Journal of Cultural Studies* 18 (3): 300–318.

Homan, Mike. 1999. *A Complete History of Fighting Dogs.* Gloucestershire: Ringpress Books.

Howe, Leo. 2005. *The Changing World of Bali: Religion, Society and Tourism.* New York and London; Routledge.

Hull, Matthew S. 2012. *Government of Paper: The Materiality of Bureaucracy in Urban Pakistan.* Berkeley: University of California Press.

Hurn, Samantha. 2012. *Humans and Other Animals: Cross-Cultural Perspectives on Human-Animal Interactions. Anthropology, Culture, and Society.* London New York: Pluto Press.

Hussain, Shafqat. 2010. "Sports-Hunting, Fairness and Colonial Identity: Collaboration and Subversion in the Northwestern Frontier Region of the British Indian." *Conservation and Society* 8 (2): 112. https://doi.org/10.4103/0972-4923.68911.

Iliopoulou, Maria A, and Rene P Rosenbaum. 2013. "Understanding Blood Sports." *Journal of Animal & Natural Resource Law* 9: 125.

Ingold, Tim. 2013. "Anthropology Beyond Humanity." *Suomen Antropologi: Journal of the Finnish Anthropological Society* 38 (3).

Jacobson, David. 1991. *Reading Ethnography*. Albany: State University of New York Press.

Jafri, Amir H. 2008. *Honour Killing: Dilemma, Ritual, Understanding*. Oxford: Oxford University Press.

Jamal, Amina. 2013. *Jamaat-e-Islami Women in Pakistan: Vanguard of a New Modernity?* New York: Syracuse University Press.

Jamal, Haroon, Amir Jahan Khan, Imran Ashraf Toor, and Naveed Amir. 2003. "Mapping the Spatial Deprivation of Pakistan." *The Pakistan Development Review* 42 (2): 91–111.

Jeffrey, Craig. 2010. *Timepass: Youth, Class, and the Politics of Waiting in India*. Stanford, California: Stanford University Press.

———. 2013. *The Global Pigeon*. Chicago and London: The University of Chicago Press.

———. 2017. "Keywords." *South Asia: Journal of South Asian Studies* 40 (2): 272–73.

Jerolmack, Colin. 2009. "Humans, Animals, and Play: Theorizing Interaction When Intersubjectivity Is Problematic." *Sociological Theory* 27 (4): 371–89.

Joseph, John. 1997. "Bear-Baiting in Pakistan." World Society for the Protection of Animal's Liberty Campaign. World Society for the Protection of Animals. 1997. www.wildlifeofpakistan.com/ResearchPapers/bear_baiting_report.pdf.

Kalof, Linda, and Carl Taylor. 2007. "The Discourse of Dog Fighting." *Humanity & Society* 31 (4): 319–33.

Katyal, Akhil. 2013. "Laundebaazi." *Interventions* 15 (4): 474–93. https://doi.org/10.1080/1369801x.2013.849417.

Kavesh, Muhammad A. 2009. "Pigeon Keepers and Their Social Status: A Case Study of Village Shahpur Phull, District Lodhran." *Anthropology*. Islamabad: Quaid-i-Azam University.

———. 2018a. "From Colony to Post-Colony: Animal Baiting and Religious Festivals in South Punjab, Pakistan." In *Colonial Transformation and Asian Religions in Modern History*, edited by David W. Kim, 10–29. Newcastle: Cambridge Scholars Publishing.

———. 2018b. "From the Passions of Kings to the Pastimes of the People: Pigeon Flying, Cockfighting, and Dogfighting in South Asia." *Pakistan Journal of Historical Studies* 3 (1): 61–83.

———. 2018c. "Beyond Cage and Leash: Human-Animal Relations in Rural Pakistan." *Anthropology*. Canberra: Australian National University.

———. 2019. "Dog Fighting: Performing Masculinity in Rural South Punjab, Pakistan." *Society & Animals* 1 (online): 1–19.

Khan, Hussain Ahmad. 2004. *Re-Thinking Punjab: The Construction of Siraiki Identity*. Lahore: Research and Publication Centre, National College of Arts.

Khan, Naveeda Ahmed. 2012. *Muslim Becoming: Aspiration and Skepticism in Pakistan*. Durham: Duke University Press.

Khan, Shaharyar M., and Ali Khan. 2013. *Cricket Cauldron: The Turbulent Politics of Sport in Pakistan*. London and New York: I. B. Tauris.

Khan, Tahira S. 2006. *Beyond Honour: A Historical Materialist Explanation of Honour Related Violence*. New York: Oxford University Press.

Kim, Claire Jean. 2015. *Dangerous Crossings: Race, Species, and Nature in a Multicultural Age*. New York: Cambridge University Press.

Kipnis, Andrew. 2011. *Governing Educational Desire: Culture, Politics, and Schooling in China*. Chicago: University of Chicago Press.

Kippen, James. 1988. *The Tabla of Lucknow: A Cultural Analysis of a Musical Tradition. Cambridge Studies in Ethnomusicology*. Cambridge: Cambridge University Press.

Kirksey, S. Eben, and Stefan Helmreich. 2010. "The Emergence of Multispecies Ethnography." *Cultural Anthropology* 25 (4): 545–76. https://doi.org/10.1111/j.1548-1360.2010.01069.x.

Knight, John. 2005. *Animals in Person: Cultural Perspectives on Human-Animal Intimacy*. New York and Oxford: Berg.

Kociejowski, Marius. 2010. *The Pigeon Wars of Damascus*. London: Biblioasis.

Kohn, Eduardo. 2007. "How Dogs Dream: Amazonian Natures and the Politics of Transspecies Engagement." *American Ethnologist* 34 (1): 3–24. https://doi.org/10.1525/ae.2007.34.1.3.

———. 2013. *How Forests Think: Toward an Anthropology beyond the Human*. Berkeley, Los Angeles, London: University of California Press.

Kumar, Nita. 1988. *The Artisans of Banaras: Popular Culture and Identity, 1880–1986*. Princeton: Princeton University Press.

———. 1989. "Work and Leisure in the Formation of Identity: Muslim Weavers in a Hindu City." In *Culture and Power in Banaras: Community, Performance, and Environment, 1800–1980*, edited by Sandria B. Freitag, 147–70. Berkeley, Los Angeles and Oxford: University of California Press.

Laucella, Pamela, C. 2010. "Michael Vick: An Analysis of Press Coverage on Federal Dogfighting Charges." *Journal of Sports Media* 5 (2): 35–76.

Lefebvre, Alain. 1999. *Kinship, Honour and Money in Rural Pakistan: Subsistence Economy and the Effects of International Migration*. Richmond: Curzon.

Lévi-Strauss, Claude. 1963. *Structural Anthropology*. New York: Basic books.

Liechty, Mark. 2003. *Suitably Modern: Making Middle-Class in a New Consumer Society*. Princeton and Oxford: Princeton University Press.

———. 2010. *Out Here in Kathmandu: Modernity on the Global Periphery*. Kathmandu: Martin Chautari Press.

Lile, Emma. 2005. "Dog Fighting." In *Encyclopedia of Traditional British Rural Sports*, edited by Tony Collins, John Martin, and Wray Vamplew. London and New York: Routledge, 100–2.

Locke, Piers, and Jane Buckingham. 2016. *Conflict, Negotiation, and Coexistence: Rethinking Human–Elephant Relations in South Asia*. New Delhi: Oxford University Press.

Loyd, Anabel. 2020. *Bahawalpur: The Kingdom That Vanished*. Penguin Random House India Private Limited.

Lumbard, Joseph E. B. 2007. "From Hubb to 'Ishq: The Development of Love in Early Sufism." *Journal of Islamic Studies* 18 (3): 345–85.

Lyell, James C. 1887. *Fancy Pigeons*. 3rd ed. London: L. Upcott Gill.

Lyon, Stephen M. 2004. *An Anthropological Analysis of Local Politics and Patronage in a Pakistani Village*. Lewiston, Queenston, Lampeter: Edwin Mellen Press.

Lyon, Stephen M, and Michael Fischer. 2006. "Anthropology and Displacement: Culture, Communication and Computers Applied to a Real World Problem." *Anthropology in Action* 13 (3): 40–53.

Maehle, Andreas-Holger. 1994. "Cruelty and Kindness to the 'Brute Creation': Stability and Change in the Ethics of the Man-Animal Relationship, 1600–1850." In *Animals and Human Society: Changing Perspectives*, edited by Aubrey Manning and James Serpell, 81–105. London and New York: Routledge.

Majumdar, Boria. 2003. "Cricket in India: Representative Playing Field to Restrictive Preserve." *Sport in Society* 6 (2–3): 169–91.

———. 2006. "Royal Cricket: Self, State, Province and Nation." *The International Journal of the History of Sport* 23 (6): 887–926.

Malinowski, Bronislaw. 1932. *Argonauts of the Western Pacific*. London: G. Routledge & sons.

Marchand, Trevor H. J. 2014. "For the Love of Masonry: Djenné Craftsmen in Turbulent Times." *Journal of African Cultural Studies* 26 (2): 155–72.

Marsden, Magnus. 2005. *Living Islam: Muslim Religious Experience in Pakistan's North-West Frontier*. Cambridge: Cambridge University Press.

Marvin, Garry. 1984. "The Cockfight in Andalusia, Spain: Images of the Truly Male." *Anthropological Quarterly* 57 (2): 60–70.

———. 1988. *Bullfight*. Oxford and New York: Basil Blackwell.

———. 2005. "Disciplined Affections: The Making of an English Pack of Foxhounds." In *Animals in Person: Cultural Perspectives on Human-Animal Intimacies*, edited by John Knight, 61–78. New York: Breg.

Mate, Bence. 2017. *Pigeon Battles of Cairo: Egypt's High-Flying Sport*. Aljazeera. www.aljazeera.com/programmes/witness/2017/09/pigeon-battles-cairo-egypt-high-flying-sport-170913073822210.html.

Mazumdar, Ranjani. 2007. *Bombay Cinema: An Archive of the City*. Minneapolis: University of Minnesota Press.

Mikhail, Alan. 2015. "A Dog-Eat-Dog Empire: Violence and Affection on the Streets of Ottoman Cairo." *Comparative Studies of South Asia, Africa and the Middle East* 35 (1): 76–95.

———. 2017. "The Moment in History When Muslims Began to See Dogs as Dirty, Impure, and Evil." India: Quartz. 2017. https://qz.com/1038116/the-moment-in-history-when-muslims-began-to-see-dogs-as-dirty-impure-and-evil/?utm_source=qzfb.

Mughal, Muhammad Aurang Zeb. 2014. "Time, Space and Social Change in Rural Pakistan: An Ethnographic Study of Jhokwala Village, Lodhran District." *Anthropology*. Durham Theses: Durham University. http://etheses.dur.ac.uk/9492/.

Mullin, Molly H. 1999. "Mirrors and Windows: Sociocultural Studies of Human-Animal Relationships." *Annual Review of Anthropology* 28: 201–24.

Mūsavī, Vālih Sayyid Muḥammad. 1788. "Kabūtar-Nāmah." British Library.

———. n.d. "Murgh-Nāmah."

Nandy, Ashis. 1983. *The Intimate Enemy: Loss and Recovery of Self Under Colonialism*. Delhi: Oxford University Press.

Narayan, Kirin. 1993. "How Native Is a 'Native' Anthropologist?" *American Anthropologist* 95 (3): 671–86.

———. 2016. *Everyday Creativity: Singing Goddesses in the Himalayan Foothills*. Chicago and London: University of Chicago Press.

Narayan, Kirin, and Muhammad A. Kavesh. 2019. "Priceless Enthusiasm: The Pursuit of Shauq in South Asia." *South Asia: Journal of South Asian Studies* 42 (2): 711–25.

Narayanan, Yamini. 2018. "Cow Protection as 'Casteised Speciesism': Sacralisation, Commercialisation and Politicisation." *South Asia: Journal of South Asian Studies* 41 (2): 331–351.

Nelson, Matthew J. 2011. *In the Shadow of Shari 'ah: Islam, Islamic Law, and Democracy in Pakistan*. London: Hurst and Co.

Nurbakhsh, Javad. 1989. *Dogs: From a Sufi Point of View*. London and New York: Khaniqahi-Nimatullahi Publications.

O'Hanlon, Rosalind. 1997. "Issues of Masculinity in North Indian History: The Bangash Nawabs of Farrukhabad." *Indian Journal of Gender Studies* 4 (1): 1–19.

Obaid-Chinoy, Sharmeen. 2015. *A Girl in the River*. HBO. www.hbo.com/documen-taries/a-girl-in-the-river-the-price-of-forgiveness/synopsis.

Oldenburg, Veena Talwar. 1984. *The Making of Colonial Lucknow 1856–1877*. Princeton: Princeton University Press.

Ortner, Sherry B. 2016. "Dark Anthropology and Its Others: Theory since the Eighties." *HAU: Journal of Ethnographic Theory* 6 (1): 47–73.

Orwell, George. 1958. *George Orwell: Selected Writings*. London & Edinburgh: Heinemann Educational Books Ltd.

Pandian, Anand S. 2001. "Predatory Care: The Imperial Hunt in Mughal and British India." *Journal of Historical Sociology* 14 (1): 79–107.

Papanek, Hanna. 1971. "Purdah in Pakistan: Seclusion and Modern Occupations for Women." *Journal of Marriage and Family* 33 (3): 517–30. https://doi.org/10.2307/349849.

Parkes, Peter. 1987. "Livestock Symbolism and Pastoral Ideology Among the Kafirs of the Hindu Kush." *Man* 22 (4): 637–60.

Pearson, M. N. 1984. "Recreation in Mughal India." *British Journal of Sports History* 1 (3): 335–50.

Pemberton, Neil. 2014. "The Rat-Catcher's Prank: Interspecies Cunningness and Scavenging in Henry Mayhew's London." *Journal of Victorian Culture* 19 (4): 520–35.

Philippon, Alix. 2011. "Sunnis Against Sunnis. The Politicization of Doctrinal Fractures in Pakistan." *The Muslim World* 101 (2): 347–68. https://doi.org/10.1111/j.1478-1913.2011.01360.x.

———. 2012. "The 'Urs of the Patron Saint of Lahore: National Popular Festival and Sacred Union Between the Pakistani State and Society?" *Social Compass* 59 (3): 289–97. https://doi.org/10.1177/0037768612449714.

Picard, Liza. 2003. *Dr. Johnson's London: Everyday Life in London 1740–1770*. London: Phoenix.

Pocock, David Francis. 1973. *Mind, Body and Wealth: A Study of Belief and Practice in an Indian Village*. Totowa, New Jersey: Rowman and Littlefield.

Price, Jennifer. 1999. *Flight Maps: Adventures with Nature in Modern America*. New York: Basic Books.

Pritchett, Frances. 1994. *Nets of Awareness: Urdu Poetry and Its Critics*. Berkeley: University of California Press.

Radford, Mike. 2001. *Animal Welfare Law in Britain: Regulation and Responsibility*. Oxford: Oxford University Press.

Rahman, Tariq. 2007. *Language and Politics in Pakistan*. New Delhi: Orient Longman.

Ray, Satyajit. 1977. *The Chess Players*. Devki Chitra Production.

Richter, William, L. 1979. "The Political Dynamics of Islamic Resurgence in Pakistan." *Asian Survey* 19 (6): 547–57.

Risley, Herbert Hope. 1915. *The People of India*. Edited by William Crooke. 2nd ed. Calcutta and Simla: Thacker, Spink & Co.

Ritter, Hellmut. 2003. *The Ocean of the Soul: Man, the World and God in the Stories of Farid al-Dīn ʿAttar*. Translated by John O'Kane. Leiden and Boston: Brill.

Ritvo, Harriet. 1994. "Animals in Nineteenth-Century Britain: Complicated Attitudes and Competing Categories." In *Animals and Human Society: Changing Perspectives*, edited by Aubrey Manning and James Serpell, 106–26. London and New York: Routledge.

Robbins, Joel. 2013. "Beyond the Suffering Subject: Toward an Anthropology of the Good." *Journal of the Royal Anthropological Institute* 19 (3): 447–62.

Rogers, Martyn. 2008. "Modernity, 'Authenticity', and Ambivalence: Subaltern Masculinities on a South Indian College Campus." *Journal of the Royal Anthropological Institute* 14 (1): 79–95.

Roseberry, William. 1982. "Balinese Cockfights and the Seduction of Anthropology." *Social Research* 49 (4): 1013–28. https://doi.org/10.2307/40971228.

Russell, Bertrand. 1932. *The Conquest of Happiness*. London: George Allen & Unwin Ltd.

Rytter, Mikkel. 2010. "In-Laws and Outlaws: Black Magic Among Pakistani Migrants in Denmark." *Journal of the Royal Anthropological Institute* 16 (1): 46–63.

Said, Edward W. 1978. *Orientalism*. London: Penguin Books.

Sakata, Hiromi Lorraine. 1983. *Music in the Mind: The Concepts of Music and Musician in Afghanistan*. Kent, OH: Kent State University Press.

Sandiford, Keith A. P. 1983. "Cricket and the Victorian Society." *Journal of Social History* 17 (2): 303–17.

Schechner, Richard. 1993. *The Future of Ritual*. New York: Routledge.

Shackle, Christopher. 2006. "Representations of ʿAttar in the West and in the East: Translations of the Mantiq al-Tayr and the Tale of Shaykh Sanan." In *Attar and the Persian Sufi Tradition: The Art of Spiritual Flight*, edited by Leonard Lewisohn and Christopher Shackle, 165–96. London and New York: IB Tauris.

Shah, Nafisa. 2016. *Honour and Violence: Gender, Power and Law in Southern Pakistan*. New York: Berghahn Books.

Shah, Rukhsana. 2018. "Animal Legislation." *Dawn*, August 7, 2018. www.dawn.com/news/1425442.

Sharar, Abdul Halim. 1975. *Lucknow: The Last Phase of an Oriental Culture*. Translated by E. S. Harcourt & Fakhir Hussain. London: Elek.

Shaw, Alison. 2000. *Kinship and Continuity: Pakistani Families in Britain*. Amsterdam: Harwood Academic Publishers.

Sims-Williams, Ursula. 2013. "Pigeon Keeping: A Popular Mughal Pastime." *Asian and African Studies Blog* (blog). London: British Library. February 23, 2013. http://britishlibrary.typepad.co.uk/asian-and-african/2013/02/pigeon-keeping-a-popular-mughal-pastime.html.

Sinha, Mrinalini. 1995. *Colonial Masculinity: The "Manly Englishman" and the "Effeminate Bengali" in the Late Nineteenth Century*. Manchester and New York: Manchester University Press.

Slobin, Mark. 1976. *Music in the Culture of Northern Afghanistan*. Wenner-Gren Foundation for Anthropological Research, Inc. Arizona: Tucson.

Smith, Page, and Charles Daniel. 1975. *The Chicken Book*. Boston: Little, Brown.

Smith, Robert. 2011. "Investigating Financial Aspects of Dog-Fighting in the UK." *Journal of Financial Crime* 18 (4): 336–46.

Soares, Benjamin, and Filippo Osella. 2009. "Islam, Politics, Anthropology." *Journal of the Royal Anthropological Institute* 15 (s1): S1–23.

Spiro, Alison M. 2005. "Najar or Bhut—Evil Eye or Ghost Affliction: Gujarati Views About Illness Causation." *Anthropology & Medicine* 12 (1): 61–73.

Srinivasan, Krithika. 2013. "The Biopolitics of Animal Being and Welfare: Dog Control and Care in the UK and India." *Transactions of the Institute of British Geographers* 38 (1): 106–19. https://doi.org/10.1111/j.1475-5661.2012.00501.x.

Stebbins, Robert A. 2015. *Serious Leisure: A Perspective for Our Time*. New Brunswick and London: Transaction Publishers.

Stoller, Paul. 1997. *Sensuous Scholarship*. Philadelphia: University of Pennsylvania Press.

Stone, Lucian. 2006. "Blessed Perplexity: The Topos of Hayrat in 'Attar's Mantiq al-Tayr." In *Attar and the Persian Sufi Tradition: The Art of Spiritual Flight*, edited by Leonard Lewisohn and Christopher Shackle, 95–111. London and New York: IB Tauris.

Storey, Charles A. 1977. *Persian Literature: A Bio-Bibliographical Survey*. Vol. 2, part 3. Leiden: E. J. Brill Ltd.

Sunstein, Cass R. 2004. "Introduction: What Are Animal Rights?" In *Animal Rights: Current Debates and New Directions*, edited by Cass R Sunstein and Martha C Nussbaum, 3–18. Oxford and New York: Oxford University Press.

Taylor, Steve. 2013. "Searching for Ontological Security: Changing Meanings of Home Amongst a Punjabi Diaspora." *Contributions to Indian Sociology* 47 (3): 395–422.

The Government of India. 1890. "The Prevention of Cruelty to Animals Act." Vol. 1 ACT XI.

The IUCN Red List of Threatened Species. 2015. "Ursus Thibetanus (Asiatic Black Bear, Himalayan Black Bear)." 2015. www.iucnredlist.org/details/22824/0. Accessed 27 October 2015.

Theodossopoulos, Dimitrios. 2005. "Care, Order and Usefulness: The Context of the Human-Animal Relationship in a Greek Island Community." In *Animals in Person*, edited by John Knight, 15–36. Oxford and New York: Berg.

Thompson, E. P. 1967. "Time, Work-Discipline, and Industrial Capitalism." *Past & Present*, 38: 56–97.

Tottoli, Roberto. 1999. "At Cock-Crow: Some Muslim Traditions About the Rooster." *Zeitschrift Fur Geschichte Und Kultur Des Islamischen Orients* 76: 139–47.

Trueman, C. N. 2015. "Pigeons and World War One." UK: The History Learning Site. 2015. www.historylearningsite.co.uk/world-war-one/the-western-front-in-world-war-one/animals-in-world-war-one/pigeons-and-world-war-one/.

Tsing, Anna Lowenhaupt. 2013. "More Than Human Sociality." In *Anthropology and Nature*, edited by Kirsten Hastrup, 27–42. New York and London: Routledge.

Turner, Victor Witter. 1967. *The Forest of Symbols: Aspects of Ndembu Ritual*. Ithaca and London: Cornell University Press.

———. 1969. *The Ritual Process: Structure and Anti-Structure*. Chicago: Aldine Pub. Co.

Varzi, Roxanne. 2006. *Warring Souls: Youth, Media, and Martyrdom in Post-Revolution Iran*. Durham and London: Duke University Press.

Veblen, Thorstein. 1912. *The Theory of the Leisure Class: An Economic Study of Institutions*. New York: The Macmillan Company.

Vorspan, Rachel. 2000. "'Rational Recreation' and the Law: The Transformation of Popular Urban Leisure in Victorian England." *McGill Law Journal* 45: 891–974.

Wacquant, Loïc J. D. 1995. "The Pugilistic Point of View: How Boxers Think and Feel about Their Trade." *Theory and Society* 24 (4): 489–535.

———. 2004. *Body and Soul: Notebooks of an Apprentice Boxer*. Oxford and New York: Oxford University Press.

Webster, John. 2005. *Animal Welfare: Limping towards Eden*. Oxford: Blackwell Publishing.

Wehr, Hans. 1979. *A Dictionary of Modern Written Arabic*. 4th ed. Urbana: Spoken Language Services, Inc.

Welch, Stuart Cary. 1985. *India: Art and Culture 1300–1900*. New York: The Metropolitan Museum of Art.

Werbner, Pnina. 2003. *Pilgrims of Love: The Anthropology of a Global Sufi Cult*. Bloomington: Indiana University Press.

Wetherell, Margaret, and Nigel Edley. 1999. "Negotiating Hegemonic Masculinity: Imaginary Positions and Psycho-Discursive Practices." *Feminism & Psychology* 9 (3): 335–356.

Whitney, Azoy, G. 1982. *Buzkashi: Game and Power in Afghanistan*. Philadelphia: University of Pennsylvania Press.

Wilkinson-Weber, Clare M. 1999. *Embroidering Lives: Women's Work and Skill in the Lucknow Embroidery Industry*. Albany: State University of New York.

Williams, Raymond. 1977. *Marxism and Literature*. Oxford and New York: Oxford University Press.

Wise, Steven M. 2002. *Drawing the Line: Science and the Case for Animal Rights*. Cambridge and Massachusetts: Perseus Books.

Wolf, Richard K. 2006. "The Poetics of 'Sufi' Practice: Drumming, Dancing, and Complex Agency at Madho Lal Husain (And Beyond)." *American Ethnologist* 33 (2): 246–68.

Woodburne, A. Steward. 1981. "The Evil Eye in South Indian Folklore." In *The Evil Eye: A Casebook*, edited by Alan Dundes, 55–65. Madison: The University of Wisconsin Press.

Zia, Afiya Shehrbano. 2019. "Can Rescue Narratives Save Lives? Honor Killing in Pakistan." *Signs: Journal of Women in Culture and Society* 44 (2): 355–378.

Index